CDL Study Guide

2024-2025

Complete Review - 1100 Questions - 39 Full Length Tests and Detailed Answer Explanations for the Commercial Driver's License Exam (All Classes)

Table of Contents

Note:

A link to the downloadable materials is on the last page.

Introduction

This book will provide you with all the information you need to excel on your CDL test. It is divided into two parts: the theory and the practice test aspects. Before you take the exam, these are some questions every driver needs to know.

What is a CDL?

CDL is an acronym for Commercial Driver's License. It is a requirement for those who will be driving heavy-duty commercial vehicles or buses on the public expressways in the United States. A CDL is proof that a driver has passed the test confirming he/she has the knowledge and skills needed for driving heavy-duty commercial vehicles.

Drivers often need endorsements for hazardous materials (hazmats), tankers, air brakes, school buses, doubles and triples and passenger vehicles.

Reading this book will prepare you for each of these endorsement tests. At the end of this book, there are practice tests, answers and explanations for each endorsement test you are likely to take.

Who Needs a CDL?

A CDL is mandatory for drivers who operate vehicles that fall under any of these categories:

- A vehicle with a gross vehicle weight rating (GVWR) of no less than 26,001 pounds.

- A vehicle that is less than 26,001 pounds GVWR and is manufactured for or used to load more than 10 passengers, including the driver.

- A combination vehicle with a 26,001 pounds or more gross combination weight rating (GCWR), and the GVWR of the vehicle attached is above 10,000 pounds.

- A vehicle that is towing another vehicle that is 10,001 pounds GVWR or more.

- A vehicle that is towing two or more vehicles or is a trailer bus.

- Any type of vehicle that has hazmat placards or contains products that are made up of selected agents according to the Code of Federal Regulations (CFR).

- A vehicle that ships hazardous waste.

- School employees who drive 15-passenger buses are required to get a passenger vehicle endorsement (PV). In a situation where the passenger seats have been removed, the bus is still considered a 15-passenger seater.

The CDL Classifications

There are three classes of combination vehicles: Classes A, B and C.

Class A: Combination vehicles

A Class A license is for commercial drivers who are operating combination vehicles of 26,001 pounds GCWR or more, with a vehicle being towed that weighs more than 10,000 pounds.

Class B: Light combinations or straight trucks

A Class B license is for commercial drivers who are operating single vehicles of 26,001 pounds GVWR or more, with a vehicle being towed that is not more than 10,000 pounds GVWR.

Class C: Single vehicles that are less than 26,001 GVWR

A Class C license is for vehicles that are manufactured for or are used to carry no fewer than 16 passengers, including the driver. This type of license is also needed if the transporting vehicle contains placarded hazardous materials.

What Is Required for Obtaining a CDL?

Although you'll have to pass seven knowledge tests, obtaining a CDL is not as difficult as it seems. These tests include a general knowledge test, an air brake test and five other endorsement exams.

The kind of commercial vehicle you intend to drive will determine the test(s) you will take. You will need to take all the tests for the relevant class license in your state.

Federal guidelines govern the types of questions on the CDL exam for every state. Although each state may have differences in regulations, they share greater similarities.

How Much Does a CDL Cost?

To get the exact cost for a CDL in the state you are planning to take the test, check with your state's Department of Motor Vehicles (DMV). Just as an example, New York State charges a compulsory driver's license fee of $164.50. Every endorsement in the state costs another $5.

How Many Tests Are There?

Expect to take seven knowledge tests. These are grouped into eight knowledge tests, which are written and skills tests. The knowledge tests are the general knowledge test, air brake test and several other specific endorsement tests.

The General Knowledge Test

H: Hazardous Materials

N: Tank Vehicle

S: School Bus

P: Passenger Vehicle

T: Double/Triple Trailer

X: Tanker and Hazmats

The Skills Test

After successfully passing the general knowledge test, you can take the skills test. You will be tested on three general skills. They are:

- Vehicle inspection skills
- Basic control skills
- Road skills (once called a driving test).

These tests are compulsory for obtaining a CDL. The test can only be taken in English. Marked or labeled vehicles are not allowed for the vehicle inspection test.

Vehicle Inspection Test

This 40-minute test focuses on your knowledge of safe driving. You will be asked to inspect a vehicle and tell the examiner what you are inspecting and why. You can take the other exam once you have successfully passed this test. You will not be charged an additional fee if you need to retake the inspection test.

Basic Control Skills Test

The examiners want to know if you are capable of handling a heavy vehicle. In this 30-minute test, they will ask you to drive the vehicle to the front, back, left, right, at specific angles and to a particular destination. Expect to be asked to recognize some markings of cones, traffic lanes, barriers and other objects. If you fail this test, you will not be able to move on to the next one until you pass. There is a retake fee.

Road Test/Drive Test

This test checks to see how well you can drive a large vehicle in different traffic conditions with a DMV route. Some of these situations are taking a left and right turn, driving through intersections and curves, driving over railroad crossings and driving on roads with single or multiple lanes, routes and streets and on highways. The examiner will determine which of these routes you will take. This test takes 45 to 60 minutes. If you fail the road test, there is a retake fee.

These are the reasons you need to take a road test:

- To obtain an original CDL
- To remove a ban or limitation on your license, especially because of the weight or size of your vehicle or the products inside
- To obtain a passenger vehicle (PV) or school bus (S) endorsement

- To renew licenses that have been expired for two or more years.

How to Check Your Skills Test Results If You Are Not in Your State

You will receive the results of your tests via electronic channels from the center or state where you took the test. Visit a DMV office in your state to complete the application process before it expires.

How Long Does the CDL Test Take?

The test lengths vary by state. But, predominantly, the general knowledge section is 60 minutes, air brakes is 25, hazardous materials is 30 and the other sections are 20 minutes each.

What Items Do You Need for the CDL Test?

Check with your state for specific requirements. Generally, you will need a valid identity card, a Social Security card, proof of passing vision and medical tests, success on the hazmats test and proof of citizenship or residence. The state agency will run a background check on you.

How Old Do You Have to Be to Obtain a CDL?

You have to be at least 18 to drive within your state. You cannot drive beyond your state borders if you are under 21 years. At 21, you may or may not be granted approval to ship hazardous materials or to have children as passengers in a school bus.

Is the Pay for Being a CDL Driver Worth It?

The salary for truck drivers ranges according to the type of product or passengers you are transporting, the destination of the product and the company you are working for. As a CDL driver, expect to earn $35,000 to $75,000 per year. Experience also determines your salary.

Chapter 1: Air Brakes

In this chapter, you will learn how brakes work to help you obtain an air brake endorsement if your vehicle needs a CDL.

This chapter covers the following topics:

• What Is an Air Brake?

• How Does an Air Brake System Work?

• Parts or Components of Air Brake Systems

• How to Use Air Brakes

• The Advantages of Air Brakes

• Maintenance/Inspection of Air Brakes

What Is an Air Brake?

Air brake systems are often found in trucks, heavy vehicles and commercial buses. Heavy vehicle drivers need air brakes to efficiently and safely stop their vehicles. An air brake system will not work if compressed or accumulated pressure is not applied to the brake component. Drivers use their legs to operate air brakes.

Air brakes are classified as high-power or heavy-brake systems. There are two types of brakes: disc brakes and drum brakes. These brakes can be further grouped into vacuum brakes, air brakes and hydraulic brakes, among others.

Driving high-power systems requires a greater source of force that is beyond human effort. Air brakes use compressed air to supply these greater forces to operate this power system. The generated force creates friction against the brakes and the tires through the brake pads.

Some mechanics prefer to call air brakes "three-in-one brakes" because the system comprises three connected braking systems:

The service brake: This component is what you use in your usual driving. It is engaged when you press and release the brake pedals.

The parking brake: When you park your vehicle, you use the parking brake.

The emergency brake: When there is a fault in the brake system, the emergency brake comes to the rescue. It uses some components of both the service brake and parking brake to stop the vehicle.

How Does an Air Brake System Work?

Whenever a driver applies pressure on the brake pedals, there is a closure of the exhaust channel. This closure leads to the opening and intake of compressed air through the air channels and into the brake chamber. When the driver returns the brake, there is an opening of the exhaust channel and a closure of the intake passage. The used air finds its way into the surrounding atmosphere.

Air brake systems are designed to have mechanical brakes for emergency purposes. An emergency brake, which is also known as an air-assisted hydraulic braking system, is used when there is a failure of air supply to the brake chamber.

When a driver tries to stop a vehicle using the brake pedal, the brake compressor kicks off as the engine gets started. The air in the atmosphere is compressed, then the compressed air travels to the air reservoir through the compressor governor. In this air reservoir, there is other compressed air that was accumulated from the earlier cycles.

If drivers want to stop a vehicle, they place drum line brakes on the wheels. The friction ignited from the impact of the brake pads against the rotation stops the vehicle. Note that whenever you apply the brake pedals, the triple valve's outlet is closed, while the inner valve opens to let the compressed air within the air reservoir travel around the brake lines.

The brake cylinder, which has an inbuilt piston, then receives the compressed air traveling within the brake line system. The compressed air on the piston forces it to adjust from its position. Pneumatic energy transforms into mechanical energy.

Brake drums are inserted down the wheel of the brake cylinder. The brake cylinder is the house of mechanical actuators, like slacks and springs. The pressure from the compressed air increases the speed of the piston, which then enlarges the mechanical actuator of the brake drum. This expanded mechanical actuator nudges the brake pads toward the outward side to create a frictionless impact on the drum lines, which are already rotating.

Parts or Components of Air Brake Systems

As a driver, you need to be familiar with the components of air brake systems. Understanding how they work will help you drive and maintain your vehicle. Also, you will be tested on how well you know the parts of air brake systems when you take the CDL exam.

Air Compressor

The air compressor is the main component of an air brake system that uses a belt drive. Its purpose is to compress air in the environment to a particular pressure and transport it to the air reservoir or storage tank.

Air Compressor Governor

This part of the air brake system controls when atmospheric air will be supplied from the compressor to the air reservoir or storage tanks. At a maximum level of about 125 PSI, the air governor makes sure there is a refrain of continual pumping of air from the compressor. However, when the pressure goes down to about 100 PSI, it permits the compressor to continue pumping air.

Storage Tank

A storage tank functions as a reservoir for atmospheric air that has been compressed under high pressure. It sustains the movement of a vehicle on a trip. Although the compressor performs a very important role in the air brake system, it needs a storage tank for sustainability. The mechanism of a storage tank makes it possible for the brakes to be applied as often as necessary.

Brake Pedal

The brake pedal is a mechanical link that uses the applied effort on the mechanism for either slowing or stopping the vehicle. The driver manually operates the brake pedal, which is inside the vehicle.

Brake Actuator

The brake actuator is connected directly to the brake pedal. Some technicians know it as a piston cylinder that produces pressure when a driver presses the brake pedal.

Safety Valve

The safety valve is attached to the air storage or reservoir tank. It makes sure the tank does not burst when there is excessive pressure from the constant air that the compressor supplies. The safety valve ejects air when the pressure gets to its maximum point.

Dirt Collector

The dirt collector precedes the triple valve. It is a tiny component that collects separated dirt from the air filters.

Brake Drum

When you apply the brake, the brake drum is engaged. You can see it attached to the tires of vehicles.

Braking Lines

Sometimes, braking lines are called supply lines. They transport compressed or stored air from the storage tank to the brake drum.

Triple Valve

The air brake is almost useless without the triple valve because it functions as a press and release mechanism for the brake. As you hit the brake pedal, the triple valve supplies the needed pressure. Then it releases or sets off the pressure when you relax the brake pedal.

Front Brake Limiting Valve

The front brake limiting valve has two modes—slippery or normal. The pressure that travels to the front brake will be limited to half if you switch to slippery mode. A limiting valve inhibits the front wheel from slipping on slick roads.

Air Filter and Air Dryer

The air filter separates dust particles from the atmospheric air. The air dryer, or inlet, dries the air, removing humidity and moisture. Every efficient brake system needs clean, dry air, or else the accumulation of dirt particles will result in faults within the brake system.

Air Tank Draining Valve

Air tanks have a draining valve that removes water and compressor oil residuals that may be in the tank. Failure to drain the water and oil may lead to faults in the air brake system if the liquids freeze during cold weather. The air tank can be drained manually or automatically. For the manual draining system, you pull a cable or make a quarter turn. The automatic draining system often has built-in electric heaters.

Alcohol Evaporator

The alcohol evaporator pours alcohol into the air brake system to avoid the consequences of trapped ice in the air brake system during cold weather.

Foundation Brakes

The wheels of a truck are operated with foundation brakes. The most common type is the S-cam, which has the shape of an S.

S-cam Brakes: Applying the brake pedal allows air into the brake chamber; the pressure from the air then makes the rod protrude. This affects the slack adjuster, which twists the shaft of the brake cam and rolls the S-cam. Then the S-cam pulls the brake shoes apart and forces them into the inner chamber of the brake drum. When you let go of the brake pedals, the wheels are free to run again because the S-cam rolls back and brings the brake shoes back from the brake drum.

Wedge Brakes: This is another type of brake, which can be operated manually and automatically. It works by having the brake chamber push the wedge against the brake shoes.

Disc Brakes: Disc brakes that use an air mechanism apply air pressure against a brake chamber in a power screw slack adjuster. The slack adjuster resembles the S-cam. The slack adjuster moves the power screw. Then, the power screw hits the rotor, which is found in a caliper brake lining pad.

Supply Pressure Gauge

Air brake systems are designed to link a pressure gauge to the air tank. If it is a double-gauge system, then expect a gauge attached to each part of the system. It may also have just one gauge that comes with dual needles. With this gauge, you can know the exact amount of pressure within the air tank.

Application Pressure Gauge

Not all vehicles have this gauge, which shows you exactly the amount of air pressure you have applied to the brakes. Remember that when you find yourself needing to apply more pressure to maintain the speed of your vehicle, the brakes may need to be replaced. A need for increased pressure may be due to mechanical faults, air leaks or brake misalignment. In general, hold on to your lower gear to slow down.

Low Air Pressure Signal

Air brake trucks and buses need this component, which tells you when your pressure supply is almost below 60 PSI, or one-half of your air compressor governor is running down on pressure. A red light or a buzzer act as a warning. Most big buses give off a warning when air pressure is at 80 to 85 PSI.

Stoplight Switch

When you turn on the switch brakes, the stoplight switch turns on the brake's lights. This serves as a warning to drivers who are behind you.

Spring Brakes

Spring brakes supply mechanical force to aid heavy vehicles in parking and emergency braking. They are engaged when there is a leak in any part of the air brake system or a reduction of air pressure to 20 to 45 PSI.

Antilock Brake System (ABS)

An antilock brake system, which is also known as ABS, helps prevent your wheels from locking. It is computerized and works with your normal brakes without reducing or increasing the effectiveness of your normal brakes. The Department of Transportation mandated that all tractors manufactured on or after March 1, 1997, should have an antilock braking system. The department also made this mandatory for all converter dollies, trucks, buses and trailers manufactured on or after March 1, 1998. Hydraulic brake trucks that weigh 1,000 pounds and were produced on or after March 1, 1999, are also required to have this component.

An ABS ensures you maintain a reasonable speed. It is not a replacement for your normal brakes. Its sole function is to stop your wheels from closing up due to too much braking.

Dual Air Brake Systems

Dual air brake systems are often used for safety purposes. Although there are two systems, they use just one set of controls. Each system comes with separate air brake parts, such as hoses, brake lines, air tanks, etc. One takes over the front axle, while the other operates the back axle. Both provide air to the vehicle. The first and second are known as the primary systems and secondary systems, respectively. Do not drive an air brake vehicle until the compressor has built up pressure to at least 100 PSI.

How to Use Air Brakes

As a commercial driver, you need to know how to use air brake vehicles. Here are some essential things you should take note of, especially in view of your CDL exam.

Normal Stops

Pull down the brake pedals. Use controlled pressure to arrive at a smooth stop. If the transmission is manual, wait until the engine RPM goes down before you press the button. Choose the starting gear after stopping.

Using ABS

Do not change how you brake when driving a truck, tractor or bus with ABS. The only exception is when you are handling a straight truck or combination vehicle that has ABS on both axles. If there is an emergency stop, use the brakes without restraint.

Emergency Stops

Sometimes, you may have another driver who is pulling over ahead of you. You can simply step on the brakes if there is a good distance between the two of you. Stop in a straight line and either apply the controlled braking or use the stab braking technique.

Controlled Braking Technique: Step hard on the brakes and do not lock the wheels. Do not make too much movement on the steering wheel. If you must make a great steering adjustment of the wheel when it is locked, then let go of the brakes and quickly return to them again.

Stab Braking Technique: Apply your brakes without stopping and let go only when the wheels are shut up. Resume brake application when the wheels reopen and start running. After the brake release, a second may pass before the wheels continue rolling. Wait until the rolling of the wheels begins before you press the brakes again. Otherwise, the truck may not be as straight as it should be. Avoid hitting the brakes hard, even in an emergency. If you do so, the wheels will not respond and the truck will skid, making it difficult to control the vehicle.

The Advantages of Air Brakes

The air brake system has many benefits.

- Air brakes work better than other types of brakes.
- The components of the air brakes are fixed where the chassis is designed.
- The compressors of air brakes can also be used for horns, tire wipers and some other functions.
- The core component of the air brake mechanism is atmospheric air, which is always available.

- Most vehicles, trucks and buses on the road use air brakes because of their effectiveness.
- Air brakes are not only for road vehicles. Railways also use them.
- With air brakes, high-pressure atmospheric air can be easily reserved in the storage tank.
- Air brakes make it easy to slow and stop heavy trucks and buses.
- When driving a vehicle with air brakes, the truck or bus can easily be controlled.
- Wear and tear of the system components is reduced.
- The hose connection of air brakes is flexible.

Maintenance/Inspection of Air Brakes

Like every other mechanical system, air brakes need maintenance and inspection. Inspecting your air brake system makes your vehicle last longer and more efficiently. Here is a basic guide to looking after an air brake system.

Engine Compartment Checks

Look for the air compressor where the air governor is situated. You can find this where the driver sits. If the air compressor uses a belt to function, make sure the belt is in perfect shape—not loose, and free of wear and tear.

Inspect the Whole System

Inspect the slack adjusters. Stop your vehicle on a level surface and make sure the wheel is not moving. You can chuck it if necessary.

During your walk-around inspection, check manual slack adjusters on S-cam brakes. Put on your gloves and pull the slack adjusters. If any of the slack adjusters go beyond one inch with the combined push rod, then you need to fix or adjust this.

Inspect the Disc Linings/Brake Drums and Hoses

Be sure that there is more than a one-half crack within the width of the friction zone. Make sure the lining is not extremely thin. No part should be faulty or missing. The air hose should be in alignment with the brake chamber. The air hose should be intact and not cut or shabby.

New Inspection Rules

In July 2015, the US government established some new rules for the inspection of air brakes. Here are some of them:

Inspection of Parking Brake: Try driving the vehicle when the forward gear is in place and the parking gear is engaged. If the vehicle does not move, that means you have a functional parking brake.

Inspection of Service Brake: Let go of the parking brake and drive the truck for a minimum of 5 mph. Hold closely to the steering wheel after slowly hitting the brake pedal. The vehicle should be stable and not move to the right or left when you apply the brakes. If the truck wobbles to either side, then there is a need for brake adjustment.

Test for Air Loss: Switch off the engine but keep the key on. Lay down the brake pedal. Look at the air gauges. In a Class A vehicle, there is not more than a loss of 4 PSI within a minute. There should not be a loss of 3 PSI in Class B vehicles.

Low Air Warning: If the air gauges are not up to 60 PSI, continue fanning the compartment of the brake pedal. Expect to see a light signaling a shortage of air. Do not stop fanning until you see the air pressure reach between 40 and 20 PSI. At this point, the brake valve will shoot out. Do not focus on the gauges when this happens. Instead, watch the valves.

Chapter 2: Hazardous Materials (Hazmats)

Every commercial driver operating vehicles that transport hazardous materials needs to study this chapter as a guide for the CDL hazardous materials test. After reading this chapter, you will be able to identify hazardous materials and the regulations guiding their maintenance, use and transportation.

This chapter covers the following topics:

- What Are Hazardous Materials?
- Classification of Hazardous Materials
- Why Hazmat Regulations Exist
- Marking, Loading and Unloading of Bulk Tanks
- Brace Containers for Safety
- Loading and Unloading of Bulk Packages
- The Responsibility of Everyone Involved in Transporting Hazmats
- Rules for Communication
- Precautionary Ways of Handling Hazmats
- Mixed Loading Prohibition
- Emergency Situations
- A Glossary of Hazardous Materials

What Are Hazardous Materials?

Hazardous materials are substances or materials that pose risks and may have adverse effects on safety, health and assets in the course of transportation. The acronym for hazardous materials is hazmat, or HM. You may see the acronym on road signs or in government policies. Examples of hazardous materials are explosives, solids, gases, flammable liquids, and other harmful materials. Hazmats pose a critical risk to drivers, passengers and citizens. Therefore, the government at all levels regulates the acquisition and use of these hazardous materials.

The Hazardous Materials Regulations (HMR) are found in parts 171-180 of title 49 of the Code of Federal Regulations. The common reference for these regulations is 49 CFR 171-180. These regulations mandate every commercial driver transporting hazardous materials display diamond-shaped warning signs, which are known as placards.

Before you are permitted to transport any toxic or hazardous materials, you must have a CDL and a hazardous materials endorsement. Before you are given the license, you must pass a test that assesses your knowledge of the regulations for transporting hazardous materials. The knowledge you need to pass these exams is covered in this section. Employees and drivers handling hazardous materials must go through special training once every three years.

Not all states have the same regulations regarding the type and quantity of hazmats that can be transported. For safety reasons, drivers are sometimes asked to drive certain routes when moving hazmat material. They are prohibited from using certain substances while driving. To further ensure maximum safety and security, the federal government often requires permits or exemptions when a heavy-duty vehicle is loaded with dangerous hazardous material, such as rocket fuel.

Classification of Hazardous Materials

Class 1: Explosives: e.g., fireworks, dynamite, ammunition

Class 2: Gases: e.g., helium, oxygen, propane

Class 3: Flammable liquids: e.g., acetone, gasoline

Class 4: Flammable solids: e.g., fuses, matches

Class 5: Oxidizers: e.g., hydrogen peroxide, ammonium nitrate, etc.

Class 6: Poisons: e.g., arsenic, pesticides, etc.

Class 7: Radioactive: e.g., plutonium, uranium, etc.

Class 8: Corrosives: e.g., battery acid, hydrochloric acid, etc.

Class 9: Miscellaneous hazardous materials: e.g., asbestos, formaldehyde, etc.

No classification: Other Regulated Material Domestic (ORM-D): e.g., asbestos, hair spray, charcoal, etc.

No classification: Combustible liquids, such as lighter fluid and fuels

Why Hazmat Regulations Exist

Hazmat regulations exist for everyone's safety. Here are the intents of hazmat regulations:

To Contain the Dangers of Hazmats

Driving hazardous materials is one of the riskiest endeavors an individual may engage in. That is why there are regulations that serve the best interest of everyone: you, the people around you, the families of those around you and the environment.

Without these regulations, transporters may not be diligent or knowledgeable enough to carefully and safely package hazardous materials. The regulations instruct drivers and companies on the safest ways for packaging, loading and unloading these materials. These regulations are also known as containment rules.

To Inform Drivers of Risks Involved

Shippers need to inform drivers about the risks involved in driving hazardous materials. Hazmat regulations instruct shippers to place a warning on every package the drivers are transporting, obtain the right shipping papers, provide information for emergency responses and display placards. These regulations inform the driver, the carrier and the shipper on the risks of hazmats.

To Ensure the Safety of Equipment and Drivers Directly in Contact with Hazardous Materials

To pass the CDL test for a hazmat endorsement, you must know:

- How to recognize hazmat materials
- How to safely load and transport materials
- How to apply placards to your vehicle.

Some drivers try to use shortcuts and break hazmat rules. Violations are punishable with imprisonment. Law enforcement officers are usually on the road to ensure rules are followed and to check if you have the necessary papers.

Marking, Loading and Unloading of Bulk Tanks

In this section, you will learn how to mark, load and unload bulk tanks.

Markings

Markings help in identification. They should be placed on portable and cargo tanks. They must highlight the shipping name on both sides. The size of the shipping name should not be less than two inches in height if the capacity of the portable tank is above 1,000 gallons or one inch if the capacity is below 1,000 gallons.

Ensure the identification number is on both sides and also on the end of any bulk tank that has a capacity of more than 1,000 gallons. If the tank holds less than 1,000 gallons, place the identification number at a parallel level. Visibility should be your priority when displaying the identification number.

Apart from portable and cargo tanks, intermediate bulk containers (IBCs) are a type of bulk package that does not need a shipping name.

Loading and Unloading Products

While loading, be circumspect when handling containers that are made up of hazmats. Avoid using materials that may puncture or damage the package. The person loading the package should stay within 25 feet of the tank.

Requirements for Loading

• Do not load or unload if you have not set the parking brake. Ensure the vehicle is static.

• Keep materials far from heat sources because most hazardous materials are flammable.

• Identify any leaks or broken containers and fix them immediately before commencing shipment. It is unlawful to transport leaking materials.

Brace Containers for Safety

The following items should be packed tightly to avoid movement or sliding of the products when they are being transported: gases, oxidizers, flammable liquids and solids, explosives, corrosives, radioactive materials and poisons.

Smoking Is Prohibited

Anything that produces fire or heat should not be brought close to hazardous materials during loading and unloading. Once you have successfully loaded and begun transit, do not open or transfer any packages until you reach the destination.

Loading and Unloading of Bulk Packages

Bulk packaging that is fixed to a truck or trailer is called a cargo tank. This is permanently attached to the vehicle when you are loading or unloading.

Bulk packaging that is not fixed to a vehicle is called a portable tank. When loading and unloading packages, the portable tank is not on the vehicle, which is static. Every portable tank should have the owner or lessee's name.

There are different kinds of cargo tanks. The most popular one is the MC306, which is meant for hazardous liquids, and the MC331, which supports gases.

The Responsibility of Everyone Involved in Transporting Hazmats

Key stakeholders for the transportation of hazardous materials are:

The Shipper

The shipper is in charge of:

• Transporting goods to different points

• Identifying a product's name; classification of hazard; identification number; correctly packaging products; labeling, marking and placing placards

• Making sure all shipping papers have been obtained.

The Carrier

Hazmat carriers are responsible for:

• Transporting products from the shipper to the final destination

• Ensuring proper documentation, marking, labeling and availability of necessary materials

• Refusing improper and illegal shipments

• Accounting to the shipper and the appropriate government agencies when accidents or unexpected incidents occur.

The Driver

The driver is responsible for:

• Ensuring that the shipper has done due diligence in identifying, marking and labeling hazardous materials

• Refusing to drive products that are leaking

• Placing necessary placards

• Transporting products quickly

• Complying with every regulation regarding the transportation of hazardous materials

• Safely keeping endorsement documentation and other vital credentials.

Rules for Communication

If there is an accident, you may be unable to communicate the types of hazardous materials that are in your vehicle. The police and other rescuers can minimize the level of risk they are at if they are aware of the kinds of hazardous materials in the vehicle.

That is why the communication rules require:

• Every shipper to highlight the types of hazardous materials in the shipment

• The inclusion of phone numbers for 24/7 response on the shipping papers

• Every carrier and driver to place hazmat shipping papers where they can be easily located in an emergency

• Every driver to store hazmat materials in a bag in the door closest to the driver or in a position where everyone can see it.

Precautionary Ways of Handling Hazmats

To prevent damage or explosion of hazardous materials:

• Switch off your engine before attempting to load or unload.

• Cut off heat sources.

• Ensure the floor lining is tight and not metallic or made of ferrous metal.

• Be extremely careful when handling explosives. Do not throw or drop the containers.

• If a package is dampened or stained with oil, do not transport it.

• Label radioactive materials as Yellow III.

• Keep Class 4 and 5 materials in secured and enclosed packages because they are harmful or reactive when wet. These materials should always be kept dry.

• Provide proper ventilation for materials that are susceptible to combustion.

• Load corrosive materials one after the other and place them upright.

• Place nitric acid below other packages.

• Make sure corrosive liquids are not close to explosives, blasting agents, oxidizers or poisonous gases.

• Create racks for storing cylinders. If there are no racks, then ensure the floor of the cargo area is flat. You may keep cylinders right side up or in a horizontal position.

• Do not keep any package labeled as poison in the driver's cab or with food materials.

• Switch off the engine before loading or unloading a flammable liquid.

• Before you park around private property, ensure the owner is aware of the dangers involved.

• Park farther than 300 feet from an open fire, a bridge, a building, a tunnel or a multitude of people. You can park only when it is very necessary and only for a short period.

• Park your vehicle in approved locations, which are also known as safe havens.

• Whenever you stop by the roadside for emergency purposes, use red lights or reflective triangles as signals. Do not use flares.

• Plan your route.

• As much as possible, avoid crowded areas.

Mixed Loading Prohibition

The hazmat rules state that some products must be loaded at different locations to avoid combustion or contamination.

Emergency Situations

When there is an incident or crash, you should:

• Ask those around to stay far from the crash site.

• Call the emergency response number immediately.

• Check the condition of your assistant or driver. Ensure everyone in the vehicle is safe.

• Keep shipping papers with you.

A Glossary of Hazardous Materials

In the field of mechanics and transportation, there are contextual terms that are used when discussing hazardous materials. They are as follows:

Bulk packaging: This is packaging that is different from a vessel. It is found in tank vehicles or trailers that have hazardous materials.

Cargo tank: This is bulk packaging and is known by three characteristics:

• It is a tank that is meant to contain tools, liquids, gases and reinforcements.

• It is a permanent or nonpermanent appendage to a vehicle. Due to its attachment to the vehicle, it can be loaded and unloaded without being disconnected from the main vehicle.

• It is not specifically designed for cylinders or any type of tanker.

Carrier: This is someone who is involved in the movement of people or products on water or land. The individual may work on a contract basis or as a private carrier.

Consignee: This is the receiver of shipment delivery. It may be a business or a person.

Division: This is a subdivision or subunit of a group or class of hazard.

EPA: This is an acronym for the U.S. Environmental Protection Agency.

FMCSR: This stands for the Federal Motor Carrier Safety Regulations.

Freight container: This is a container that can be used multiple times. It can contain 64 cubic feet volume of gas or liquid. It is fabricated to allow the intact raising of its contents when loading or unloading for transport.

Fuel tank: This tank is different from a cargo tank and is used to ship flammable liquids and gases that power or fuel the vehicle to which it is fixed. It may sometimes serve other tools that are in the vehicle.

Gross weight/mass: This is the total mass or weight of the container and the product.

Hazard class: This is a classification of hazardous materials based on their description and form.

Hazardous materials: This is a solid, liquid or gaseous substance that the secretary of transportation has designated to be a threat to the safety, health and assets of citizens. Hazardous materials include water pollutants, toxic waste and substances at volatile temperatures.

Intermediate bulk container (IBC): This is portable packaging that can be rigid or susceptible to adjustments. It is made to take the place of a portable tank or cylinder.

Limited quantity: This is the maximum quantity for which a hazardous material may have a labeling exception.

Markings: These are the name, description, caution, identification number or specifications on the outside body of packaging.

Mixture: This is the compounding of two or more chemical elements.

Non-bulk packaging: This is smaller than the packaging size of bulk packaging. For liquids, the maximum capacity is 459 liters, which equates to 119 gallons. For solids, the maximum mass is below 400 kg, which is equivalent to 882 pounds. For liquids, the highest capacity is 450 liters or 119 gallons.

NOS: This means Not Otherwise Specified.

Outage: This is also known as the ullage. It is the quantity in which a packaging goes below its full capacity. It's often stated in volume.

Portable tank: This is a type of bulk packaging that has a maximum capacity of 1,000 pounds. It does not consist of a tank car or cargo tank.

PSI: This means pounds per square inch.

PSIA: This means pounds per square inch absolute.

RQ: This means reportable quantity.

RSPA: This is the Research and Special Programs Administration.

Shipper's certification: This is a statement in which the shipper asserts on a shipping paper that the shipment is prepared and packaged in compliance with the law.

Shipping paper: This is also known as a manifest, shipping order or common bill of lading. It contains information about the transported products.

Technical name: This is the chemical, botanical or scientific name of a substance.

Transport vehicle: This can be a trailer, van, truck, tank car or tractor that is used to transport attached cargo from the point of loading to the final destination.

Chapter 3: Combination Vehicles

This chapter is for all drivers who intend to drive combination vehicles, such as tractor-trailers, truck trailers, doubles and triples. Make sure you read Chapter 4 of this book if you need more information on doubles and triples.

This chapter covers the following topics:

• What It Takes to Drive Combination Vehicles

• The Air Brakes of Combination Vehicles

• Procedures for Coupling and Uncoupling Combinations

• Inspecting Combination Vehicles

What It Takes to Drive Combination Vehicles

Combination vehicles differ from other commercial vehicles in certain ways. They weigh more, are bigger and demand greater driving dexterity. Combination vehicles include straight trucks, trailers, doubles, triples and tractor-trailers.

The knowledge you use in driving a car is not the same knowledge you use when driving combination vehicles. Therefore, you need to pay attention to the details in this section.

Preventing Rollover Risks

Rollovers cause half of all crashes. Piling too much cargo into a combination vehicle will affect its stability. The truck will lose balance and can easily tilt off the road. The possibility of an empty truck rolling over is minimal compared to a fully loaded truck.

To retain stability when driving, drive slowly when taking a curve and ensure the cargo is not stacked too high. Do not change lanes abruptly. Fast driving causes rollovers.

Steering gently helps prevent a crack-the-whip tilting of your trailer. Be gentle when pulling over and never make an unexpected, sudden move.

Do not drive too close to other vehicles. Maintain a three-second distance of 30 feet in front of your truck. Look ahead of you as you drive.

Early Braking

Speed is dangerous. Take it slow with your trailer. Combination vehicles take longer to stop when they are empty than when they are loaded to capacity. Light cargo makes the brakes' suspension stiff. Maintain a reasonable following distance between your vehicle and the vehicle in front of you so that you can easily apply the brakes early.

Preventing Skidding

Look at the mirrors to observe that your trailer is stable. Check your mirrors any time you apply your brakes to be sure the trailer is in the right position. It is not easy to stop accidental skidding or jackknifing when the trailer has already swerved out of the lane.

Also, do not use the brakes if you want to regain traction.

Using a Turn-Wide Technique

The front and rear wheels are often in different positions when you are going around a corner. This is called cheating or off-tracking. Expect longer trailers to have more cheating than shorter ones. To come back to alignment, steer the front very wide along the corner so that the back end does not roll up on the curb, pedestrians or anything else that is in the way. The rear of your vehicle should not be far from the curb, so other vehicles will not be able to come in front of you on the right.

The Air Brakes of Combination Vehicles

Go through Chapter 1 of this book before you read this section. In addition to the components of air brake systems, combination vehicles have their own unique system.

Trailer Hand Valve

Some people call the trailer hand valve the Johnson bar or the trolley valve. It is used to test the brakes and should not be engaged in driving because it may later skid. It is better to use the foot brake when driving because it transfers air to every part of the air brake system and reduces the possibility of skidding.

Also, the hand valve should not be involved when parking because it might lead to air leaks. Instead, use parking brakes to park your vehicle and wheel chocks to prevent its movement.

Tractor Protection Control Valve

This component maintains the air in the air brake system if there are leaks in the trailer or if it breaks. The air supply control valve controls the tractor protection valve. If the air pressure is around 20 to 45 PSI, then the tractor protection valve will close up without manual application. This closure prevents air from escaping the air brake system. This mechanism causes the air to move through the emergency line.

Trailer Air Supply Control

You can identify this component by the red, eight-sided knob found in modern vehicles. It controls the air supply control. When you push the knob, air is supplied; when you pull it out, the air supply is shut off.

Air Lines

All combination vehicles come with service and emergency lines. They operate to and from both vehicles.

Service Air Line: Also called the signal or control line, it transfers air around the air brake system. The service line pressure is based on the effort you exert on the brake or hand valve. The service line is linked to the relay valves, which facilitate the quicker application of trailer brakes.

Emergency Air Line: This is also called the supply line. It has two basic functions—to supply air to the air tank and to control the emergency brakes. Service lines are often blue, while emergency or supply lines are often red.

Glad Hands or Hose Couplers

Glad hands help couple the service and emergency lines in the truck/tractor together with the trailer. The couplers come with rubber seals that help prevent air from escaping. Always ensure the rubber seal is clean before you connect the trailer with the truck. Join both seals and the couplers in a 90-degree position.

Some combination vehicles are designed with dead ends, which are also called dummy couplers. These are connected with the hose when they are idle. This keeps water and dirt particles away from the couplers and the two air lines. Dummy couplers should be used when the emergency and service lines are not connected to the trailer. In the absence of dummy couplers, you can lock both glad hands.

Shut-Off Valves

These are also known as cut-out cocks and are for the supply and service air lines located behind towing trailers. These valves allow the closure of the air line when there is no towing trailer.

Procedures for Coupling and Uncoupling Combinations

Understanding the procedures for coupling and uncoupling is necessary for the safety and longevity of your vehicle.

How to Couple Tractors/Semitrailers

Step 1: Inspect the fifth wheel.

- Look for parts that have been destroyed or which are missing.
- Inspect the security and completion of your tractor.
- Check to see that the fifth wheel is well greased to avoid frictional steering problems.
- Inspect the fifth wheel position and set it in the proper coupling position.
- Lock up the fifth wheel if it is sliding.
- Check if the kingpin is damaged or out of shape.

Step 2: Observe the area and chock the wheels.

• Make sure the area around the vehicle is clear.

• Ensure the spring brake is on and the wheel is chocked.

• Check to make sure the cargo is intact and not susceptible to movement.

Step 3: Set the position.

• Always make sure the trailer is directly behind the tractor. It should not be different from any angle.

• Use your side mirrors to inspect the position of your tractor.

Step 4: Slowly back up.

• Keep backing up until the trailer is against the fifth wheel.

• Do not hit the trailer.

Step 5: Ensure the security of the tractor.

• Turn on the parking brake.

• Set the transmission to neutral.

Step 6: Observe the height of your trailer.

• The trailer should not be so low that the tractor can bring the trailer up when it is backed to the trailer. A very low trailer may make the tractor hit the trailer and destroy the nose. A very high trailer may make coupling difficult.

• Ensure the alignment of the kingpin.

Step 7: Couple the air lines to the trailer.

• Make sure the glad hands are intact and link the emergency line to the trailer's emergency glad hand.

• Connect the servicing air line to the trailer's service glad hand.

• Ensure the protection of air lines so that they are not crushed when the tractor is taking the back of the trailer.

Step 8: Provide air to the trailer.

• Press the air supply button or switch the valve control from emergency to normal mode so that air can go through the air brake system.

• Wait for the air pressure to become normal.

• Look to see if there are crossed lines in the air brake system.

• Switch off the engine so you can hear the sound of the brakes

• Listen to the sound as you press and let go of the trailer brakes. You should hear brake movement when the brake is applied and the release of air when the brake is released.

• Inspect the pressure gauge to see if there is any air loss and ensure the air pressure is moderate.

• Start the engine when you are certain the brakes are functioning effectively.

Step 9: Lock up the trailer brakes.

• Pull the air supply button or change the valve control from normal to emergency position.

Step 10: Backing the tractor to the trailer.

• Apply the reverse gear, which is at the lowest position.

• Go slow when backing the tractor under a trailer. Stop once there is an interlock between the fifth wheel and the kingpin.

Step 11: Check the security connection.

• Slightly lift the landing gear from the base.

• Ensure the trailer is coupled with the tractor.

Step 12: Secure the vehicle.

• Set the transmission to neutral.

• Turn on the parking brake.

• Switch off the engine and take the key out.

Step 13: Supervise the coupling.

• Get a flashlight.

• Ensure no gap exists between the lower and upper fifth wheels.

• On the underside of the trailer, check the back of the fifth wheel, ensuring there is a closure of the fifth wheel jaw along the kingpin shank.

• Make sure that the lever for locks is in the lock position and the safety latch is over it.

• To prevent accidents, fix any aspect of coupling before you start driving.

Step 14: Plug the electrical cords in and inspect both air lines.

Step 15: Lift the landing gear/front trailer support.

• Start from the lowest point. When the landing gear has become weightless, increase the gear range.

• Keep on lifting the landing gear until it is at the top and not halfway to avoid it being trapped on railroad tracks.

• Ensure the safety of the crank after the landing gear has been raised.

• Ensure there is enough distance between the landing gear and end part of the tractor frame if the trailer is completely on the tractor. This is to make sure that the landing gear does not hit the tractor in a sharp turn.

• There should also be a reasonable distance between the trailer nose and the top of the tractor tires.

Step 16: Remove the wheel chocks and keep them in a safe place.

How to Uncouple Tractors/Semitrailers

It is easy to uncouple tractors if you follow this procedure:

Step 1: Set the rig in the right position.

• Check the parking area and ensure it can take the weight and size of the trailer. Not all surfaces are good supports for the weight of trailers.

• Establish an alignment between the tractor and the trailer.

Step 2: Reduce the pressure against the locking jaw.

• Switch off the air supply to lock the brakes.

• Back up slowly to reduce the pressure on the fifth wheel.

• Turn on the trailer's parking brakes when the tractor is impacting on the kingpin.

Step 3: Chock the wheels to prevent movement.

Step 4: Put down the landing gear.

• Bring down the landing gear until it touches the ground.

• After the landing gear touches the ground and the trailer has been loaded, turn the crank slowly. The trailer's additional weight will be cut off.

• The fifth wheel should be left down so that it will not be difficult to unlatch it.

Step 5: Disconnect the electrical cable from the air lines.

• First set apart the air lines and the trailer, then fix the air line glad hands with the dummy couplers, which are at the rear of the vehicle.

• Hang the electrical cable to keep it from gathering moisture.

• Use only accepted or supporter lines to avoid unexpected incidents on the road.

Step 6: Uncouple the fifth wheel.

• Lift the release handle and place it in the open position.

• Stay away from the back of the tractor wheels to avoid injury if the vehicle moves.

Step 7: Pull the tractor away from the trailer.

• Keep pulling the tractor until the fifth wheel is pulled from the base of the trailer.

• Pause if the trailer is on the frame of the tractor. It will stop the trailer from tumbling to the ground if the landing gear breaks.

Step 8: Ensure the security of the tractor.

• Use the parking brake.

• Put the transmission in neutral.

Step 9: Check trailer supports.

• Ensure the ground is appropriate for parking.

• Check that the landing gear is intact.

Step 10: Clear the trailer from the tractor.

• Release the parking brakes.

• Inspect the area and continue driving the tractor until it is clear of the trailer.

Inspecting Combination Vehicles

Apart from the inspection procedures discussed in Chapter 1, there are some other things you should check your combination vehicle for.

Coupling system areas: Follow the guidelines given in the earlier section of this chapter.

Landing gear: Make sure it is fully raised, there are no missing components or leaks and everything else is secured.

Make sure **air** is flowing to all parts of the trailer.

Inspect the **tractor protection valve.**

Make sure the trailer's **emergency brake** and **service brake** are working well.

Chapter 4: Doubles and Triples

This chapter covers the scope of doubles and triples on the CDL knowledge test, including the following topics:

- How to Pull Trailers: Doubles and Triples
- Parking Doubles and Triples
- Checking the Air Brakes of Doubles and Triples
- The Process of Coupling and Uncoupling Doubles and Triples
- Inspecting/Maintaining Doubles and Triples
- The Components of Doubles and Triples

How to Pull Trailers: Doubles and Triples

You need to be very meticulous when pulling doubles and triples. Unexpected incidents can occur because no other commercial vehicles are as volatile as these kinds of trailers.

Preventing a Rollover of Doubles and Triples

Just like combination vehicles, rollovers can be prevented if you are very gentle with the steering. Drive slowly on ramps, corners and curves. Driving singles and combinations is very different from driving doubles and triples. A reasonable speed for the former may be life-threatening for the latter.

Watching out for the Crack-the Whip Effect

The crack-the-whip effect makes doubles and triples more unstable and more likely to topple over than combination vehicles. The trailer at the end may topple if you do not drive slowly.

Being Vigilant and Keeping Your Eyes on the Road

Your mind and physical body need to be alert. Do not let down your guard and always keep your eyes on the road. Watch out for speed bumps, traffic, curves and other unforeseen factors that may pose an obstacle to the smooth movement of double and triple trailers.

Maintaining a Good Distance or Space

Doubles and triples occupy more space on the road than other vehicles do. They are longer and cannot be stopped abruptly. It takes a few seconds before they can be fully stopped. Therefore, do not drive too close to other vehicles. Maintain a good distance. Do not be in a hurry to cross traffic. Confirm if the space is long enough for your vehicle. Before you change lanes, ensure that no vehicle is next to you.

Driving in Unsafe Conditions

If weather conditions are not ideal for driving, you should either wait for them to clear or be extra careful on the road. Slippery roads and fog often lead to accidents. A double and triple vehicle requires greater handling skills in these adverse conditions.

Parking Doubles and Triples

Park your vehicle in an area that you can easily pull out from. Assess the parking space to see if it is a fit for your vehicle.

Checking the Air Brakes of Doubles and Triples

Checking the air brakes on doubles and triples is similar to checking them on combination vehicles. However, there are slight differences and additional steps you should follow:

• Ensure there is a free and maximum flow of air around the trailers. Engage the trailer's parking brake and/or stop the moving vehicle through chocking or stalling of the wheel. After waiting for the air pressure to become normal, press the red switch, which supplies air to the emergency line. If you want the air to get to the service line, apply the trailer's hand brake.

Move on to the rig at the back and open the closed emergency line valve. If you listen attentively, you will hear the sound of escaping air. This shows that you have supplied power to the system.

You can now shut the valve of the emergency line and free up the service line to ensure that all the parts of the trailer have adequate air pressure.

Now, you can shut off the valve. Remember, you should hear the sound of air coming out of the air lines. If you do not, go back to the shut-off valves placed on the doubles/triples and the converter dolly and open them. Air must go to all the areas if you want the brakes to function effectively.

Inspect the tractor control/protection valve. Power up the air brake system of your doubles and triples by pressing the air supply button after mounting up the air pressure to normal. Then turn down the engine. Intermittently press and release the brake pedal to limit the quantity of air pressure in the air tank. The tractor protection valve control, also known as the trailer supply control, pops up and changes its position to emergency when the range of the air pressure is between 20 and 40 PSI or according to the manufacturer's manual.

If the tractor protection valve malfunctions, there could be a brake leak in which the air transported from the tractor comes out. This makes the emergency brakes pop out. If care is not taken, there might be a loss of control.

Inspect the trailer's emergency brakes. Power up the air brake system and make sure there is nothing to stop the trailer from rolling. Then stop and bring out the air supply control. Or, you can simply set it in the emergency position. Slowly pull out the trailer and the tractor to ensure the trailer emergency is switched on.

Inspect the trailer's service brakes. Ensure the air pressure is normal, then open the parking brakes, gently drive the vehicle to the front and press the trailer's brakes using the trolley valve or hand control. You will observe that the brakes begin to come up. This means there is a functional integration of the trailer brakes. Use the hand valve to inspect the trailer's brakes, but use your foot pedal if it is for a basic function. The foot pedal will mount air on the service brakes and every wheel of the system.

The Process of Coupling and Uncoupling Doubles and Triples

Correct coupling and uncoupling is necessary for safe travel.

How to Couple Twin or Double Trailers

Ensure the Safety of the Rear Trailer

The rear trailer may not have spring brakes. If that is the situation, bring the tractor and the trailer closer, integrate the emergency line, charge up the trailer's air storage tank, then set the emergency line apart. These settings will put up the emergency brakes after correctly adjusting the slack adjusters. If you doubt the brakes' functionality, chock the wheels.

Place the heavier semitrailer before other lighter trailers, which should be at the back.

A converter gear on a dolly is a coupling tool of one or two axles and a fifth wheel, whereby a semitrailer can be connected to the rear of a tractor-trailer combination and establish a double-bottom rig.

The second trailer should be behind the converter dolly. Open the air tank to let go of the dolly brakes. You can use the parking brake to do this if there are spring brakes.

Use your hand to wheel the dolly into alignment with the position of the kingpin if the gap is not much. You can lift the converter dolly through the front semitrailer and the tractor.

- Set the combination very close to the converter dolly. The dolly should be at the back of the front semitrailer. You can then join both.

- Lock the pintle hook.

- Protect the dolly in a lifted position.

- Keep the nose of the rear semitrailer closer to the dolly.

- Bring down the dolly support.

- Remove the dolly from the trailer that is right behind the tractor.

Integrate the Converter Dolly with the Front Trailer

- Place the front semitrailer ahead of the dolly tongue.

• Connect the front trailer with the dolly.

• Lock the pintle hook.

• Safely set the converter gear in a lifted position.

Connect the Converter Dolly to the Rear Trailer

• Lock the trailer brakes and chock the wheels.

• Set up the height of the trailer. It should be a little bit below the middle of the fifth wheel so that the trailer can rise a bit higher when the dolly is pushed from below.

• Set the converter dolly beneath the rear trailer.

• Safely set the landing gear in a lifted position and above the ground to avoid damage when the trailer is driven forward.

• Pull on the pin of the rear semitrailer to check the coupling.

• Inspect the coupling. Ensure there is no gap in the lower and upper fifth wheels.

• Set up safety chains, light cords and air hoses.

• Lock up the converter dolly's air tank. Close up the valves that are behind the rear trailer.

• Lock up the valves that are behind the front trailer.

• Lift the landing gear without restraints.

• Press the air supply button to charge up the trailer's brakes. Then open the locked-up emergency line to see if there is air at the back of the rear trailer. The absence of air means the brakes will not work because there is a mechanical fault somewhere.

How to Uncouple Twin or Double Trailers

Uncouple the Rear Trailer

• Set the rig firmly on clear ground.

• Press the parking brake so that the rig is stable.

• If your vehicle does not have spring brakes, lock up the rear trailer wheels.

• Release the rear trailer's landing gear to take the weight or pressure off the dolly.

• Lock up the shut-off behind the front trailer.

• Remove all connections between the air dolly and the electrical lines.

• Let go of the dolly brake and the latch of the fifth wheel.

• Gently move the tractor. The front trailer will be pulled first, and the dolly will move out of the back of the semitrailer.

How to Uncouple a Converter Dolly

• Bring down the dolly's landing gear.

• Remove the safety chains.

• Press the spring brakes of the converter or chock the wheels.

• Unlatch the pintle hook, which is on the front semitrailer.

• Gently drive away from the dolly.

If the dolly is beneath the rear trailer, do not open the pintle hook. Expect the dolly bar to fling up. This may result in injury and cause difficulty in coupling again.

How to Couple and Uncouple Triple Trailers

To couple the first semitrailer or tractor with the second or third trailers:

• Apply the process for joining tractors to semitrailers.

• Set the converter dolly in its place and apply the procedures for joining doubles in coupling the first and second trailers.

• When you want to uncouple a triple trailer rig, pull out the dolly to disjoint the third trailer from the other trailers and tractor.

Inspecting/Maintaining Doubles and Triples

Couple the System Parts

• Inspect the lower or bottom fifth wheel.

• Ensure that it is safely placed on the frame.

• Check to see if there are missing, leaking or faulty components.

• Sufficiently grease the system area.

• Identify and close up gaps that exist between the top and bottom fifth wheels.

• Close up jaws that are within the shank.

• Sit down comfortably and let go of the arms while the safety lock is engaged.

• Inspect the upper or top fifth wheel.

• Ensure the glide plate is safely placed on the frame.

• Ensure the kingpin does not malfunction.

• Fix the electric lines and air lines to the trailer.

• Ensure the electrical cord is tightly plugged and safe.

• Correctly connect the air lines to the glad hands. Ensure there are no air leaks and that there is proper and sufficient slack.

• Ensure that there is no fault in any of the air lines.

• Ensure the fifth wheel slide is not faulty or missing any components. Sufficiently grease the fifth wheel. Keep all locking pins available and locked. To ensure that the frame of the tractor does not destroy the landing gear when it turns, make sure the fifth wheel is not set too far.

Assessing the Landing Gear

• Keep the gear totally lifted. Check to see if there are missing or faulty parts that need replacement or repair.

• Securely maintain the crank handle.

• Watch out for air leaks.

The Components of Doubles and Triples

• Keep the valves closed. Those at the back of the front trailer should be open; the valves at the last trailer's rear should be closed. The air tank's valve system and the air tank of the converter dolly should be closed.

• Ensure all air lines are well fixed and the glad hands are securely connected.

• Secure the spare tire placed on the converter gear.

• Ensure proper locking of the pintle.

• Provide security for the safety chains.

• Properly connect light cords to the sockets on the trailers.

Chapter 5: Tanker License

All you need to know about tanker licenses and vehicles is discussed in this chapter.

You will need a tanker license if your tanker is transporting gases or liquids that are in a fixed cargo. The cargo tank capacity may be 119 gallons or above, or it can be a portable tank that has a capacity of 1,000 gallons or more. Every vehicle that requires a Class A or Class B CDL needs a tanker license.

This chapter covers the following topics:

- How to Drive Tank Vehicles
- Rules for Driving Tankers Safely
- Inspecting Tankers

No matter how solid your vehicle may seem, do not load, unload or drive it until you have inspected or checked the condition of the vehicle. You are performing this inspection to keep the liquid or gaseous contents safe and to ensure safe driving.

How to Drive Tank Vehicles

You need special knowledge and skills to drive a tanker vehicle because there is great liquid turbulence, and an accentuated center of gravity involved.

Greater Center of Gravity

A greater part of the weight of the contents of a tanker vehicle is raised high and almost off the road. Therefore, the top of the vehicle becomes weighty and is likely to fall over. It is easier for liquid loads to roll down. Rollovers are most likely to occur on curves. Therefore, curves should be taken slowly.

Danger of Liquid Overflow or Surge

When the tank is nearly full, expect liquid movement that will lead to a liquid surge. The liquid movement will negatively affect your ability to control the vehicle. When you are stopping, the liquid will move forward and backward. If a strong wave hits the rear of the tanker, expect the tanker to tilt toward the direction of the wave. If the vehicle is parked on slippery ground, the wave can move the vehicle in the direction of the wave. Every tanker driver needs greater experience for better control.

Bulkheads

Bulkheads are used to partition liquid tanks into smaller units. Any time you are loading or unloading, make sure the weight is equally distributed among partitions. Avoid overloading the front or the back of the tank with excess weight.

Baffled Tanks

These are liquid tanks with bulkheads. They have holes that allow the passage of liquid. You can have better control of a front, back or sideways liquid surge when you use baffles. Avoid liquid surge as much as possible because it causes the contents to run over.

Smoothbore or Unbaffled Tankers

This type of tanker does not have any tool to reduce liquid surge. If you use unbaffled tankers to transport liquid contents, expect a very high movement of the tanker. Unbaffled tankers are better used for transporting food products, like milk. For sanitation purposes, baffles should not be used to transport food products because it is not easy to clean the inside of the tank.

Outage or Room for Expansion

Avoid loading a tanker to its brim. Liquids expand when warm and need space, hence the need for an outage. All liquids do not expand at the same rate. Therefore, you need to be cautious of the type of liquid and the outage you are leaving. Know about the outage rate before you load liquid contents.

Quantity of Liquid

If the liquid is very dense (acid), a completely full tank may be beyond the lawful weight limitation. Therefore, if the liquid is dense, do not totally fill the tank. The three main determinants of the quantity of liquid you should load are:

- The potential for expansion during transportation
- The liquid's weight
- Legal weight limitations.

Rules for Driving Tankers Safely

Tankers are not easy to drive. The following are the rules for driving tankers safely:

Drive Smoothly

The liquid surge and high center of gravity require smooth transport. Slow down, change lanes carefully and stop smoothly.

Take Control of the Liquid Surge

Use consistent pressure on the brakes. Take it slow when you are stopping. Keep a good following distance between you and the vehicle in front. Increase the stopping distance if the road is wet. Always apply the brakes ahead of the stopping area. Apply stab or controlled braking to stop quickly and prevent a crash. Do not turn the steering wheel too quickly while braking because the vehicle may slide. It may be more difficult to stop an empty tank than a full one.

Take Curves Slowly

Do not drive too fast on curved roads. You need to take extra care to avoid turbulence.

Avoid Skids

Too much acceleration or braking may make the vehicle skid. Skidding may make your tanker jackknife if you do not take quick action to stop it.

Inspecting Tankers

In addition to studying this book, go through your tanker's specific operator's manual. There are unique components of tank vehicles that you should check for:

Leaks

Leaks are the most essential problem you need to watch out for. Walk around and check for leaks beneath or in any part of the vehicle. It is very risky to load and drive a vehicle with leaks. It is also illegal. If you are caught, authorities will stop you from continuing the drive, and you may be mandated to clean up any spills along the route you have already driven. To be thorough with your inspection for leaks:

- Inspect the body or outer part of the tank for leaks.

- Inspect the intake, discharge and shut-off valves. Fix the valves in the most appropriate places before you start loading, unloading or driving the tanker.

- Inspect all the connections and pipes.

- Inspect the vents and manhole covers. Ensure there are available gaskets that keep the covers closed. Ensure the vents are free from particles or obstructive substances.

Equipment with Unique Uses

These are also called special purpose tools. If your tanker contains any of the tools listed below, make sure they are working correctly.

- Grounding wires

- Bonding wires

- Vapor recovery package

- Fire extinguisher.

Do not drive a tanker with valves and manhole covers that are not closed.

Always perform a practical inspection of all necessary equipment. Make sure it is working before you take it with you.

Chapter 6: Passenger Transport

If you intend to transport passengers, you need a CDL if the vehicle carries 16 or more passengers, counting the driver. You also need a passenger endorsement. You will need to apply the knowledge gained in Chapter 1 and this book's FAQ to excel in the knowledge test.

This chapter covers the following topics:

- Inspecting the Vehicle Before the Trip
- Trip Procedures: How to Load and Start Every Trip
- Transporting Hazardous Materials
- Arriving at Bus Stops or the Final Destination
- Road Practices: What to Do While Transporting Passengers
- Inspecting the Vehicle After the Trip
- Prohibited Practices: What You Shouldn't Do Before, During or After Transporting Passengers

Inspecting the Vehicle Before the Trip

Safety is your priority when it comes to driving a passenger vehicle. Do not allow riders to board the vehicle if they are carrying common hazards, such as car batteries or gasoline. You must ensure every object and person you are driving is safe before loading and embarking on a trip. Go through the inspection report, which the last driver made, to see what is lacking. Refuse to sign if necessary repairs or replacements have not been made. Your signature implies that you are aware of the current state of the vehicle and have approved of it. Here are the major aspects you need to check.

Vehicle Parts

Before you drive, ensure the following components are in order:

- The parking brake
- The service brakes, which may include air hose couplings on vehicles with a semitrailer or trailer
- The reflectors and the lights
- The steering system
- The horn
- The tires (ensure the front tires are not re-grooved or recapped)

- The wipers
- The mirrors
- Coupling tools
- The wheels and tires
- The emergency kits.

Entrance/Exit Doors and Access Panels

Inspect the emergency exits and access panels and shut them if they are open. The access panels are basically for restroom services, engine, baggage, etc. Do not drive until you are certain of safety.

Vehicle's Interior

Before you load and drive your vehicle, check the interior and ascertain that it is safe for everyone to enter. Here are some vital interior areas to check:

- Handholds and railing
- Floor covering
- Signal tools
- Emergency exit handles
- Driver and passenger seats (no seat should be unfastened or loosened)
- Emergency windows and door (they should be closed)
- Emergency exit sign (this should be bold and clearly written on the emergency door; always keep the emergency light on at night)
- Roof hatches (you may partially open them for fresh air, but do not leave them open)
- Fire extinguisher (ensure this is available and functional)

Also ensure there are spare parts for the vehicle, such as tires and electrical fuses.

Seat Belts

Inspect the seat belts. Always use yours and advise your passengers to use theirs.

Trip Procedures: How to Load and Start Every Trip

- No rider or passenger should be allowed to place baggage by the aisle or doorway. Clear any obstruction that may cause passengers to stumble. Show riders where to keep their baggage and help them stow luggage when necessary.

- Make sure it is easy to get from seats to the exit and from the outside to the inside.
- Inform riders that they can exit through the window or the closest door if there is an emergency.

Transporting Hazardous Materials

Ensure you are aware of what every piece of cargo or baggage contains. Remember, you are not allowed to transport hazardous materials if you do not have a hazmat license or endorsement.

As discussed in Chapter 2, hazardous materials are threats to the health, property and safety of the people who are around the vehicle.

If you have a license to transport hazardous materials, then ensure you clearly mark and label every hazardous item appropriately.

Buses are allowed to transport small-arms ammunition that is ORM-D (other regulated materials for domestic transport only), emergency supplies for hospitals and drugs. However, as a passenger driver, you are not permitted to carry any of the following:

- Poison gas, irritating material or liquid poison
- Solid poisons that weigh more than 100 pounds
- Explosives near or on passengers' seats
- Radioactive materials near or on passengers' seats
- Hazardous materials with a total weight of more than 500 pounds and less than 100 pounds of any of the classifications listed in Chapter 2
- Gasoline and battery cells.

The Standee Line

Vehicles that are made to allow standing should have a marked line of two inches. Called the standee line, this shows passengers where they can and cannot stand. Note that no passenger should ever stand behind the driver's seat.

Arriving at Bus Stops or the Final Destination

When you are almost at a bus stop or final destination:

- Announce the destination.
- State why you are about to stop.
- Announce the time of departure before and/or after every stop.

• Announce the number of the vehicle.

• Remind passengers to take all their belongings.

• If you are driving a bus for hire or a charter, ensure no one steps into the bus until it is time to move. This avoids theft and vandalism.

Road Practices: What to Do While Transporting Passengers

These are things to do when you are transporting passengers:

• Supervise passengers.

Before you begin a trip, clearly state the rules against smoking, drinking, sexual intercourse, loud music and other prohibited activities. Be sure that passengers are aware of and understand each of the rules. If possible, invite and answer questions before you begin the trip.

Your responsibility as a driver continues while on the road. Use the mirrors to supervise what the passengers are doing while you are driving. You may have to caution offenders and remind them of the rules. Correct those whose heads or arms are outside the window. You must be vigilant.

• Be vigilant at stops.

Some passengers may fall when entering or exiting the bus. Always remind them to be cautious of their steps to avoid stumbling. Before you start driving, wait for every rider to be seated comfortably to avoid injury.

If you encounter a belligerent rider, carefully handle the situation to ensure everyone is safe. Do not be in a hurry to drop off the belligerent passenger until it is safe to do so. You may have to wait until the next stop.

How to Handle Common Bus Crashes

Most bus crashes occur at intersections. Do not depend on traffic signs alone. Use discretion when driving. Be observant enough to notice incoming vehicles, poles or objects in the way. Sometimes you may not be the driver at fault; you still need to prepare for unexpected incidents caused by reckless drivers.

Be sure there is an available gap when you need to join a lane or re-enter traffic after a stop. Do not expect other drivers to give you space after you have started pulling out. Never assume. Wait for the gap to be created before you swerve to fill in the space.

Speed on curves: Slippery roads and speed often result in the loss of lives and property. Most roads are designed for an acceptable speed. However, the speed limit for a car may not be appropriate for a bus, so you must be careful with the speed of your vehicle. Slow driving is far better than high speeds on curves.

When Stopping at Railroad/Highway Crossings (RR Crossing)

• Apply the brake when your bus is about 15 to 50 feet from the railroad crossing.

• Watch out for trains by listening and looking. If possible, step out of the cab to check for incoming trains.

• Do not be too quick to jump into a lane after a train passes. Wait to be sure there are no other trains approaching from another track.

• When crossing a track, always keep a steady gear when driving a manual transmission bus.

• Slow down to watch for other vehicles crossing the tracks.

What to Do at Drawbridges

If the drawbridge does not have a traffic control attendant or green signal light, apply your brakes. You do not need to stop, but you do have to slow down. Make sure you are at least 50 feet from the drawbridge. Check to ensure the drawbridge is fully closed up.

Inspecting the Vehicle After the Trip

Do not wait until the end of a work day before you inspect your bus. Inspection should be done after every trip. Interstate carriers often require an inspection report for every trip. The driver should highlight in the report any damages that need repair. If there are no damages, the driver should state so in the report.

Passengers often destroy vehicle parts, such as seat belts, windows, seats, etc. It is your duty to check for these damages and make a report so that the mechanics can repair the damages before the next shift.

Prohibited Practices: What You Should Not Do Before, During or After Transporting Passengers

- Do not refuel a bus if there are riders in the vehicle, unless it is very necessary.
- Be focused while driving.
- Do not get carried away with chitchat or anything that takes your mind off the road or your passengers.
- Except when the safety of your passengers is threatened, never push or tow a faulty bus when the passengers are on board. If the area is not safe, push the bus to the nearest safe zone and unload the passengers so that you can continue pushing or towing.

Chapter 7: School Bus Endorsement

This chapter explains all you need to know about driving a school bus so that you can pass the CDL test.

This chapter covers the following topics:

• Student Management in Driving

• The Process of Loading and Unloading

• Danger Zones for School Bus Drivers

• How to Use Mirrors When Driving Schoolchildren

• Railroad Crossings

• Procedures for Exiting and Evacuating in Emergency Situations

• Safety Tips for School Bus Drivers

Student Management in Driving

To be an effective school bus driver, you must understand the ways of children and how to manage them while driving.

While loading and unloading, your priority is focusing on what is happening outside. Do not ever take your gaze from the road to settle a dispute on the bus.

Here are a few safety tips you should observe for effective student management while driving:

• Observe the school rules and regulations regarding discipline before, during and after transit.

• Stop the bus as soon as you find a safe parking area.

• Turn off the vehicle and take your key out before you step out of the vehicle.

• Speak firmly but politely to the offender(s). Do not express anger in your tone or facial expressions, but be serious so that those involved will follow instructions.

• Remind the offender(s) of the rules and regulations.

• To resolve any conflict, you may switch both parties from their seats. Place them where they cannot see or reach each other. You may decide to instruct the offender to sit closer to you for easier monitoring.

• On no occasion should you remove a student from the bus unless you have arrived at the official bus stop. You may call the school administration or the police to inform them so that they can pick up the student if the offense is very serious. Just remember to comply with discipline protocols.

The Process of Loading and Unloading

Passengers on school buses are less likely to have accidents than students who are boarding or getting off a bus. Therefore, you need to be fully aware of every necessary step in loading and unloading students. In this section, you will learn the best and safest loading and unloading procedures, which every school bus driver needs to know. These procedures will help avoid injuries or fatal incidents.

When Approaching a Bus Stop

Schools have approved routes and bus stops. If you notice a new route or bus stop that should be used, do not use the new route or make a stop until you have officially written to the school district for approval.

When approaching a designated bus stop, be very cautious. This area demands vigilance and skills. You will have to use the mirrors, lights, signal arm and crossing control arm.

- Decrease the speed of your vehicle.
- Check for passersby, objects and traffic while stopping. Even after you have stopped, check again.
- Check the mirrors repeatedly.
- Switch on the warning lights 5 to 10 seconds or 100 to 500 feet before you arrive at the bus stop.
- Switch on the right-side indicator three to five seconds before you are about to pull over.
- Stay very close to the side of the driving lane.
- Stop the bus 10 feet from the bus stop. Doing this will make the students walk toward the bus so you will be able to see them fully.
- Set the transmission in neutral or park. It should be in neutral if park does not exist.
- You can open the service door to switch on the red lights if the traffic is far from the school bus.
- Wait for traffic to completely clear before you open the door and ask the students to start coming in.

Procedures for Loading at a Bus Stop

- Students are instructed to wait for the school bus to arrive at a specific bus stop. This location should be where the driver can see the students as they are approaching the bus stop.
- No students should enter the bus until the driver has signaled them to come in.
- Always use the mirrors to monitor your environment. They are essential safety tools you must continually use.

• Take a count of the students and make sure no one misses boarding the bus. You can learn the names of the students. If the number of students on board does not tally with the count or the expected number, ask present students for the whereabouts of the missing student(s). If you are not satisfied with the answer, safely park the bus and check around or beneath the bus.

• Ask the students not to rush in. Instead, they should be in a single-file line and use the handrails. Turn on the dome light of the loading area if it is dark.

• Avoid being hasty when loading. Ensure everyone is seated and ready to be transported before embarking on the trip.

• Use the side and overhead mirrors to identify if someone is running to catch the bus and be patient as the student comes in.

• Close the door, apply the transmission and switch off the alternating flashing lights.

• Use the mirrors to ensure it is safe to move.

• Wait until there is no traffic.

• When you are certain it is safe to drive, embark on the route.

Procedures for Unloading When You Are En Route

• Stop safely at the official unloading sites.

• Ask the students not to leave their seats until you tell them to do so.

• Then follow the procedures for loading and unloading.

Procedures for Unloading on a Road

Not all students will get off at a stop that is directly in front of their homes. Some will have to cross the street that is directly in front of the school bus. Observe these procedures if the student(s) will cross the road.

• Get out of the bus and walk about 10 feet from it to where you can see the students crossing.

• Stay on the right side of the road, where you can easily view the students' feet.

At the edge of the road, the students should:

• Look in all directions and ensure the road is clear for crossing.

• Observe if the bus's red lights are still flashing.

• Not cross until you have given them the signal to cross after looking in all directions.

Procedures for Unloading at School

The following guidelines are for all states:

- Switch off the ignition.

- Take your key with you when leaving your seat.

- Ask the students to remain sitting until they are told otherwise.

- Stand where you can oversee the unloading.

- Ask the students to step down in an organized manner, row by row.

- When it seems like they are all out, go back in the bus to be sure that there are not any students who have not unloaded.

- Make sure no students reenter the bus.

- After parking the bus and unloading the students, count all the students to be sure none are missing. If any are missing, check around to find them.

- After you have accounted for each student, shut the bus door, put on your seat belt, turn on the engine, run the transmission, let go of the parking brake, switch off the alternating flashing lights, switch on the left turn signal and look at the mirrors again.

- After ensuring that there is no traffic, you can leave the unloading base.

Danger Zones for School Bus Drivers

The danger zone of a school bus is an area where schoolchildren are likely to be badly injured if they are hit by their bus or another bus. The danger zones of your bus may go as far as 30 feet away from the front bumper. However, the first 10 feet are often more dangerous. Ten feet behind, 10 feet from the left and 10 feet from the right are also the most dangerous. The left zone of a bus is called a danger zone because of vehicles passing by.

How to Use Mirrors When Driving Schoolchildren

Adjust the Mirror

To avoid accidents, you need to properly adjust the mirrors. Ensure all the mirrors give you the visual field you need.

The Outside Flat Mirrors on the Left and Right Sides of Your Vehicle

You will see these mirrors at the left and right corners in front of the school bus. Use them to check for oncoming vehicles and activities going on behind the vehicle.

A blind spot is directly in front and under each mirror, as well as behind the rear bumper. The distance of the blind spot is about 50 to 150 feet; sometimes it goes as far as 400 feet based on the size of the school bus.

Properly adjusting the mirrors will help you have a view of:

- Approximately 200 feet, or the length of four buses at the back of your bus

- The sides of the bus

- The rear tires where they touch the ground.

The Outside Convex Mirrors on the Left and Right Sides of Your Vehicle

You will see these mirrors directly under the flat mirrors outside your vehicle. Use them to get a panoramic view of what is happening on the left and right sides of your outside environment. Like the flat mirrors, they give you a view of the students on the side, traffic and other activities. These mirrors will not give you the exact distance and size of objects and people.

Properly adjusting the mirrors will help you have a view of:

• Every side of the bus

• The front part of the back tires where they touch the ground

• The traffic lane.

The Outside Crossover Mirrors on the Left and Right Sides of Your Vehicle

You will see these mirrors on the left and right front sides of your vehicle. Use them for a good view of danger zones in front of and outside of the bus. It is hard to see these zones with your eyes alone. These mirrors will give you a proper view of the front wheel and the service door areas. These mirrors will not give you the exact distance and size of objects and people.

Properly adjusting the mirrors will help you have a view of:

• Every zone in front of the bus, starting from the front bumper to where you can have a direct vision. Use both your eyes and the mirrors for the best view of all sides of the bus.

• All the front tires where they do not touch the ground.

Occasionally view the flat, convex and crossover mirrors sequentially to keep everyone safe in the danger zones.

The Inside Overhead Rearview Mirror

You will see this mirror suspended above the windshield in the driver's area. Use it to check what is going on inside the bus. Instead of turning your head and taking your concentration off the road, use the overhead rearview mirror to monitor the inside of the bus. The only visibility limitation is when the bus has a glass rear emergency door.

Expect to have a blind spot that consists of the back of the driver's seat and the back of the bumper extending to 400 feet behind the bus. To have a view of this blind spot, use the side mirrors. This way you can monitor the incoming traffic around the blind spot area.

Properly positioning this mirror will help you have a view of:

- The peak of the back window that is on top of the mirror
- Every student inside the bus.

Railroad Crossings

There are two different types of crossings—passive and active.

Passive crossings: There is no traffic control machine at this type of crossing. You have to stop and apply the normal traffic procedures. Identify if imminent trains are coming before you cross over. Passive crossings come with round yellow warning signs and markings for easy recognition.

Active crossings: These crossings typically have a traffic control machine that regulates traffic. These active crossing devices are red lights that may or may not have bells.

Signals and Devices

Warning Sign: This sign is round, black and yellow. It is attached to railroad crossings. When you see a warning sign, slow down and stop for an imminent train.

Pavement Markings: These consist of a letter X and RR painted on the road. There is a no-passing mark attached.

If you see a white line painted on the pavement directly in front of the track roads, it means you should stop.

Crossbuck Sign: This can be used in place of the pavement markings. When you see it, stop.

Red Light Signals: These are flashing lights on a crossbuck sign. They mean you should stop because of an oncoming train.

Procedures for Exiting and Evacuating in Emergency Situations

Always expect emergencies to occur. They can come up at any time, to anyone and in any place, and they can be fatal. An emergency can come in the form of a fire incident, a crash, an unmovable bus in the middle of the road, a health crisis, etc. In these critical moments, what you do before, during and after can make the difference between life and death.

Plan Ahead

You can appoint two older and more responsible students to guard the emergency exits. Show them how to help students evacuate in emergencies. One additional student should be appointed to guide other students to a safety zone. If there are no older students, teach all the students what to do in an emergency. Show them how to handle the exit doors. Also, ensure you have your safety kit secured.

Identify the Necessity for Evacuation

This is the most important step in dealing with an emergency. The best way to respond is to keep the students on board to ensure everyone is safe and prevent a bad situation from becoming worse. You must be certain of the appropriate solution before you start evacuating the students.

Understand that there is a difference between the danger of fire and an actual fire and between an actual fuel spillage and the smell of fuel. Examine the facts before you expose the students to more danger by bringing them out into the open.

There are several situations where evacuation is compulsory:

- There is an actual fire or an obvious threat of fire.
- The bus suddenly stops close to or on a railroad crossing.
- A forthcoming collision requires evacuation.
- There is spillage of hazardous materials.

Evacuating to a Safe Zone

Here are tips for utilizing a safe zone:

- It should not be less than 100 feet from the road where traffic is incoming. This safe distance will protect the students from any debris that may result from a collision.
- Guide the students toward the direction of air when there is a fire or spillage of hazardous materials.
- Take students far from the railroad tracks.
- If there is an imminent tornado and the bus is in its path, lead the students out of the bus and guide them to a nearby ditch or a channel if there is no nearby building for shelter. Ask the students to lie facedown on the ground with their hands over their heads. The safe zone should be far from the bus and other vehicles. Choose a site that is safe from flash floods.

General Evacuation Tips

- Decide if you are going to use the front, the back or the side door for evacuation. Identify if the safest route for evacuation is the window or the roof. Do not hurry the students toward the door. Analyze the situation and choose the best exit.
- Set the brakes, switch off the engines and switch on the vehicle's hazard warning lights.

• Call the dispatch office to inform them of the emergency and location and to tell them the kind of assistance you need. If there is no signal or the radio is not functional, see if a passing motorist can help you. If none of these is available, send two mature students to go for help.

• Ensure the evacuation is orderly.

• Be careful about moving a student who has sustained a neck or spinal injury. This is to avoid exacerbating the injury. Leave injured students where they are unless they are in grave and immediate danger.

• Go around and check to confirm that every student has been removed from the danger zone. Take a count of all students and check that they are all safe. Then secure all emergency tools.

• Secure the safe zone and set up warning devices.

Safety Tips for School Bus Drivers

Using Strobe Lights

Many school buses have white strobe lights on the roof. You should use these lights to increase visibility in the dark.

Driving in Stormy Weather

It is more difficult to keep control of the steering wheel when a strong wind is hitting your vehicle. This kind of wind may threaten to push your bus off the road or send it off course. Here are some tips for controlling your vehicle in high winds:

• Grip the steering wheel strongly. No matter how hard the wind works against you, keep a firm grip because the safety of every student on board depends on you.

• Anticipate the direction of the wind and swerve to avoid the wind hitting you directly.

• Slow down to reduce the impact of the wind. You may have to pull over and wait for the wind to die down. Do not worry about a delay. Safety is always your priority.

• If you are unsure of other steps to take, notify your dispatcher of the situation and ask for help if needed.

Backing up a School Bus

Backing up a school bus should only be a last resort if there is no better and safer way to drive your vehicle forward. On no occasion should you back up a vehicle when the students are still outside it. Reverse gear is risky and might lead to a collision with a pedestrian or another vehicle.

If you think backing up is the only way out, observe the following procedures:

- Ask someone to be a lookout. Their role is to inform you if you are about to hit someone or something. Their duty does not include showing you how to back out. Your lookout should be observant and loud enough to give you warnings.

- Ask your students to be quiet so that you can hear your lookout and the sound of approaching vehicles.

- Use all the mirrors.

- Keep it slow. Do not hurry. An unexpected person or object may zoom in and result in a collision.

- If you cannot find a lookout, park the vehicle, switch it off, take the keys and step out of the bus. Go to the back of the bus to be sure the road is clear.

FAQ

Safety Rules for Driving All Commercial Vehicles

To ensure you understand how to safely drive commercial vehicles without jeopardizing your life or the lives of passengers and others on the road, you will be tested on safety in your CDL test.

The following are areas to look out for:

Inspecting Your Vehicle

An inspection helps you identify vehicular faults that should be resolved before the beginning of a journey.

The three basic procedures for inspecting a vehicle are:

- Pre-trip inspection
- During the trip inspection
- After the trip inspection.

Basic Control of Your Vehicle

Controlling your vehicle requires four skills:

- Accelerating
- Steering
- Stopping
- Backing up safely.

Adjusting Gears

How you shift your gears matters a lot for safe driving. You can use any of the following to adjust gears:

- Manual Transmissions
- Multi-Speed and Auxiliary Transmissions
- Automatic Transmissions
- Retarders.

Looking While Driving

Expert drivers try to look no less than 12 to 15 seconds ahead of them. If the speed is high, this is calculated as a quarter mile. If the speed is low, it is calculated as one block.

Communicating While Driving

Communication can be a source of distraction to your driving. Therefore, you should not make calls while driving. Communicate instructions only to passengers. Turn down the volume of the radio when communicating.

Appropriately Control Speed

Unchecked speed often leads to accidents. Always check your speed and drive according to the surface and environmental conditions.

Space Management

Ensure there is sufficient space on all sides of the vehicle. When there is an unexpected event, you can quickly think of what to do and manage the situation using the space between you and other vehicles and the pedestrian lane.

Remaining Aware of Hazards

Hazards were discussed in Chapter 2. You need to be vigilant about hazardous materials in your vehicle as well as in other vehicles around you.

Managing Distractions While on the Road

Keep your mind on the road. Limit conversations. Carefully select the radio programs you listen to and watch out for distracted drivers.

Dealing with Road Rage or Impatient Drivers

Impatient, aggressive drivers may not consider the rights of other drivers.

Road rage is defined as using a vehicle for the purpose of harming others.

The best way to deal with this kind of attitude is to be patient, consider others, swallow your pride, refuse to react to offensive attitudes and call the police if necessary.

Drive Safely in the Dark

Driving at night can be challenging. To drive safely at night, follow these basic steps:

- Get enough rest before a trip so you can stay alert on the road. It is advisable to sleep if you are feeling drowsy. Do not drive when you have not slept enough. Do not wear sunglasses at night.

- Do not blind other drivers with your headlights. Reduce the brightness of your lights when another vehicle is 500 or fewer feet in front of you.

• Avoid being blinded by other vehicles. When you see vehicles coming toward you, do not look directly at their headlights. Instead, look to the right. If the inside of your vehicle is very bright, this will affect your ability to see outside. Switch off the lights inside and leave on only the device lights that help you to monitor the gauges and the brake.

• When you feel sleepy, take a break. Park your vehicle in a safe site and take a nap before you continue your journey.

How to Drive Safely When in Fog/Mist

Fog is often not predicted. And when it occurs on the highway, it may be very dangerous. It affects your ability to see what is coming in your direction or what is in your path. The safest thing to do when the road is foggy is to park your vehicle at a safe distance and wait for the fog to lift.

If you must enter the fog, do not rush in. Turn on your fog lights and be wary of other drivers who may not have turned on theirs. You can use highway reflectors to guide you along the curves.

How to Drive Safely in Winter

Check your vehicle's functionality before you start driving in winter. Pay attention to the following:

• **Temperature Gauge and Coolant/Antifreeze Levels** – Ensure levels are sufficient to keep the system from freezing. Use a coolant tester to make sure it is full. If it is not, fill it up.

• **Heating Equipment** – You should not drive your vehicle in winter if you are not certain the defrosters and heaters are in good working condition.

• **Wipers and Washers** – Inspect the blades of the windshield wipers and be sure they are working well.

• **Tires** – The depth of the treads of your tires should be about 4/32 on front tires and 2/32 on other tires.

• **Tire Chains** – Take tire chains with you and learn to fix them when they need repair.

• **Windows and Mirrors** – Remove ice from the mirrors before starting your journey.

• **Radiator Shutters and Winter Front** – Clean the ice from the shutters and do not close up the winter front extremely tight. The engine may overheat if the winter front is closed tight and the radiator shutters freeze.

When driving in winter, drive slowly on slippery surfaces. If it is too slippery to drive, stop in a safe spot.

How to Drive Safely When It Is Very Hot

When the weather is very hot, you are advised to check the tires every 100 miles or every two hours. High temperatures often increase the level of air pressure. If you discover that your tires are becoming too hot, stop and wait for them to cool off to prevent an explosion. Do not try to reduce the air in the tires, because this might reduce the necessary amount of pressure.

Always have engine oil available. If you have a gauge that tells you the temperature of the oil in your vehicle, always ensure the temperature is within the right range.

How to Drive Safely in Mountainous Terrain

When driving in the mountains, you must be aware of gravity, which greatly affects vehicles. It will slow you down when you are driving uphill. If the grade is very steep, then the grade will be very long. When you are carrying a heavy load, make use of lower gears to go up mountains or hills.

When you are coming down a steep downgrade, gravity accelerates your speed. Manually gear to an appropriate speed. Slow down so that your brakes can keep your vehicle steady without overheating. Extreme heat leads to wear and tear of brakes.

It is best to plan and make inquiries about the route before you embark on mountain driving. Talk to other drivers who have driven that route before.

How to Drive Safely in Emergency Situations

When an accident occurs, take these three basic steps:

- Secure the area.
- Call emergency numbers.
- Take care of those who are injured.

Fire Control Measures

To prevent fires while driving:

- Perform a pre-trip and en route inspection of the vehicle parts.
- Handle the brakes correctly.
- Have fire extinguishing equipment ready for use.
- Always monitor the condition of your vehicle.

The Risk of Using Alcohol and Hard Drugs Before or During Driving

Alcohol and hard drugs affect your vision, perception, muscle coordination and reaction time. If you must arrive at your destination after drinking alcohol or taking hard drugs, ask someone to help you do the job.

1 General Knowledge Test

1. Why should you apply the brakes while driving on the steep downgrade of a mountain?
A. It is a key aspect of braking.
B. It makes driving enjoyable.
C. It supplements or supports the brake effect the engine has on the vehicle.
D. It stops brake fading.

2. You need __________ when backing up.
A. A greater hold on the steering wheel
B. To apply the parking brake
C. To ask the students to get out of the vehicle first
D. To load every student first

3. Which of the following is not a type of vehicle inspection procedure?
A. Pre-trip inspection
B. During the trip inspection
C. After the trip inspection
D. Stop and drive inspection

4. In a traffic situation, you should do which of the following when entering the road or changing lanes?
A. Expect vehicles to give you space for entry.
B. Wait for enough of a gap.
C. Rush in because other drivers may not allow you to enter.
D. None of the above.

5. Which of the following is the best way to communicate with other drivers when there is a hazard or obstacle that requires reduced speed?
A. Park your vehicle immediately.
B. Communicate with the emergency flashers to warn drivers behind you.
C. Wave your hands out the window.
D. Run over the obstacles.

6. ABS does which of the following?
A. Helps decrease the brake pressure
B. Helps increase the pressure in the tires
C. Gives your vehicle greater weight.
D. Prevents your wheels from locking up

7. Before you enter a curve, you should ________.
A. Increase the speed.
B. Move the gear to the right downshift.
C. Shift your transmission to neutral
D. Upshift to the right.

8. Your vehicle may__________ if there is water on the road.
A. Get hot
B. Hydroplane
C. Speed up
D. Skyrocket

9. Escape ramps are built into the downgrades of steep mountains to ___________.
A. Safely stop runaway vehicles
B. Help drivers with faulty brakes
C. Make it easier to drive
D. Increase the speed of vehicles

10. What should you expect when your brakes get too hot?
A. Your steering wheel will malfunction.
B. Your ABS will become faster.
C. The brakes will develop faults.
D. The brake drum will become bigger.

11. What is the function of ABS?
A. To increase stopping
B. To drive faster
C. To reduce the function of the brakes
D. None of the above

12. You should inspect your vehicle to ensure which of the following?
A. The safety of everyone before driving
B. The safety of everyone when driving
C. The safety of everyone after driving.
D. All of the above

13. When on a trip, you should always ____________.
A. Keep your eyes on the gauges.
B. Use your five senses for observation.
C. Inspect the tires and wheels when you temporarily stop.
D. All of the above.

14. Why should you take the ignition keys with you before you start the pre-trip inspection?
A. The keys are needed for inspection.
B. You are keeping them safe.
C. It lets people know you are the driver of the vehicle.
D. None of the above.

15. You should__________ if you realize you will arrive late at your destination.
A. Increase your speed
B. Not stop driving, no matter what
C. Accept the situation and arrive late
D. None of the above

16. After the start of a trip, every driver should do a cargo inspection within the first _________.
A. 10 miles
B. 25 miles
C. 50 miles
D. 100 miles

17. What happens when your vehicle moves around a corner?
A. The rear wheels will move to another path.
B. The rear wheels will not move to another path.
C. You have to move to the left side.
D. None of the above.

18. What is the effect of traveling twice as fast?
A. A two times increase in stopping distance
B. A three times increase in stopping distance
C. A four times increase in stopping distance
D. A six times increase in stopping distance

19. What should you do when you are tailgated?
A. Increase your following gap.
B. Switch on the taillights.
C. Do not allow the vehicle behind to overtake you.
D. Increase your speed.

20. Which of the following affects stopping distance?
A. Weather
B. Time
C. Speed
D. All of the above

21. You should stop at railroad crossings if:
A. Your vehicle has placards.
B. There is not enough fuel in the tank.
C. You are not sure of the speed.
D. All of the above.

22. What do you do when it is difficult to switch to the next gear if you are using double clutching?
A. Switch to neutral.
B. Reduce the speed of the engine.
C. Let go of the clutch.
D. All of the above.

23. When there is an engine fire, you should ___________.
A. Turn on the engine.
B. Let the fire burn.
C. Open up the hood.
D. None of the above.

24. Too much or very little air pressure ____________.
A. Is necessary for the tires
B. Will make the trip last longer
C. Makes the tires long-lasting
D. None of the above

25. A very hot, unventilated vehicle can lead to which of the following?
A. Being energetic
B. Sleepiness
C. Proper concentration
D. Nausea

26. What should you do when you feel groggy?
A. Continue driving.
B. Take medication.
C. Stop and take a nap.
D. Have a cup of coffee.

27. If there is a problem with the vehicle, what should you do to prevent a crash?
A. Park your vehicle in a safe spot and resolve any problem.
B. Hurry to get to your destination.
C. Inspect the vehicle only after the trip.
D. None of the above.

28. Add ______to the size of a school bus to know the size of its safety area.
A. 10 feet
B. 15 feet
C. 20 feet
D. 50 feet

29. You should park where ____________.
A. You will be able to back out.
B. There are tracks.
C. You will be able to move forward when leaving.
D. Other drivers cannot enter.

30. When going down a hill, you should ___________.
A. Go hard on the brakes.
B. Not go too fast and use a controllable gear.
C. Accelerate and switch to neutral.
D. Keep an eye out for emergency exits.

31. When driving on a very slippery road, you should do which of the following?
A. Stop driving.
B. Park in a safe place.
C. Find another route.
D. All of the above.

32. When should you turn off your retarder?
A. If the road is wet or filled with snow
B. When there is heavy traffic
C. If you need to keep to a strict schedule
D. When you need to increase your speed

33. Which of the following is false regarding a spare tire?
A. It should be inflated.
B. It should be the size of the normal tire.
C. It should not be damaged.
D. It should not be used in hot weather.

34. What sign should you check for to prevent a fire when driving in very hot weather?
A. Heat
B. Skidding of the wheels
C. Position of the brakes
D. Number of passengers in the vehicle

35. Federal and state inspectors may inspect your vehicle. What is the penalty for driving an unsafe vehicle?
A. The vehicle will be auctioned off.
B. You will be jailed for a minimum of six months.
C. The vehicle will be deemed out of service.
D. You will be fined $150.

36. Which of the following is true?
A. A faulty ABS means a truck has no brakes.
B. A faulty ABS means a truck will depend solely on normal brakes.
C. A faulty ABS means there will be an increase in the speed of your vehicle.
D. None of the above.

37. Which of the following is not a type of brake?
A. Wheel brake
B. Service brake
C. Emergency brake
D. Parking brake

38. What is the rule for using turn signals?
A. Signal early.
B. Signal continuously.
C. Cancel the signal.
D. All of the above.

39. If coolant is added to a system when there is no overflow tank, you should shut off the engine and do which of the following?
A. Wait for the engine to cool.
B. Slowly switch the radiator.
C. Wait for the pressure to be released.
D. All of the above.

40. Rims that have been repaired through welding ____________.
A. Can only be used for the front tires
B. Are usable with cargo
C. Are not safe
D. Can only be used for the rear tires

41. Which of the following is true when the temperature makes ice start melting?
A. Driving is advisable.
B. The road will start to crack.
C. The temperature of the vehicle will decrease.
D. The road becomes very slippery.

42. How many minutes does it take for air pressure to build up in a pre-1975 single-air system?
A. Two
B. Five
C. Three
D. Seven

43. Which of the following is an important responsibility during a trip?
A. Using your senses to detect problems
B. Monitoring the gauges to be sure there is no trouble
C. Inspecting very vital components at every stop
D. All of the above

44. Which of the following is not true when inspecting the engine compartment?
A. The wheels should be chocked.
B. The parking brake should be off.
C. You should lift the hood.
D. The door should be open.

45. A lot of heavy-duty vehicles use what kind of air brakes?
A. Single
B. Dual
C. Triple
D. Plastic

46. Which of the following is true when you are about to turn?
A. You should check the traffic to your left.
B. You should remove your hands from the wheel.
C. You should change gears.
D. You should not move the vehicle into oncoming traffic.

47. You are on a wet surface and your vehicle is hydroplaning. What should you do?
A. Use the brakes to reduce your speed.
B. Apply the parking brake.
C. Stay on the accelerator.
D. None of the above.

48. Which of the following is correct if you need to stop before turning?
A. Set the front wheels facing forward.
B. Stop abruptly with skidding.
C. Stop after the crosswalk.
D. Allow the vehicle to roll.

49. What should you do if your bus is stalled on railroad tracks?
A. Get everyone out of the bus immediately.
B. Get everyone out of the bus after a mechanic arrives.
C. Keep everyone together and close to the tracks.
D. None of the above.

50. The CDL exam will test your capability to handle a heavy vehicle, including being able to identify which of the following?

A. Markings of cones

B. Traffic lanes

C. Barriers

D. All of the above

2 Air Brakes Test

1. Air brake systems are often found in which of the following?

A. Trucks

B. Heavy vehicles

C. Commercial buses

D. All of the above

2. You need __________ to drive high-power systems.

A. Human effort

B. Mechanical effort

C. Concentration

D. All of the above

3. What are the types of air brakes?

A. Mechanical brake, automatic brake and transmission brake

B. Starting brake, driving brake and parking brake

C. Parking brake, service brake and emergency brake

D. Neutral brake, drum brake and forward brake

4. How does an air brake system function?

A. The exhaust channel closes when there is pressure on the brake pedal.

B. Closure results in the passage of air along the air channel and the brake chamber.

C. The used air is released.

D. All of the above.

5. Which of the following should you do if you want to stop your vehicle?

A. Place the drum line brakes on the wheels.

B. Release the drum line brakes from the wheels.

C. Use the service and emergency brakes even if there is no emergency.

D. None of the above.

6. Which of the following parts of the air brake system is in charge of compressing air from the atmosphere into the reservoir?

A. The air system

B. The air tank

C. The drum brake

D. The air compressor

7. When the compressed pressure of the air brake system is high, what happens?
A. The tire may burst.
B. The atmospheric air enters the storage tank.
C. There could be an explosion.
D. You turn down the electronics in your vehicle to reduce the pressure.

8. The storage tank is ____________.
A. Also known as the reservoir
B. The most important part of the air brake system
C. Useful only for combination vehicles
D. All of the above

9. Which of the following statements is true about brake pedals?
A. They are a nonessential part of the air brake system.
B. They are used for stopping the vehicle in transition.
C. They operate automatically.
D. All of the above.

10. Which of the following parts of an air brake system is meant to ensure safety in the storage tank?
A. The fire extinguisher
B. The safety valve
C. The safety net
D. The safety gear

11. The brake drum does which of the following?
A. Transfers air from one brake to another
B. Stores air after the storage tank is full
C. Carries out the action after brake application
D. All of the above

12. When does the front brake limiting valve become very necessary?
A. When driving in summer
B. When driving on a slippery road
C. When avoiding traffic
D. None of the above

13. The air dryer does which of the following?
A. Separates particles
B. Dries the air
C. Needs the air filter component
D. All of the above

14. How many types of air tank draining systems are there?
A. Two
B. Three
C. Four
D. Five

15. Applying the brake pedal will do which of the following?
A. Prevent air from passing into the brake chamber
B. Dislocate the slack adjuster
C. Keep the S-cam back
D. None of the above

16. When _________ is operating, the brake chamber pushes the wedge in between the rear of the brake shoes.
A. The wedge brake
B. The shoe brake
C. The disc brake
D. The S-cam brake

17. Air brake systems connect the supply pressure gauge to which of the following?
A. The air tank
B. The power switch
C. The brake pedal
D. The steering wheel

18. Many big buses give off a warning when air pressure is at ___________.
A. 40 to 45 PSI
B. 60 to 65 PSI
C. 70 to 75 PSI
D. 80 to 85 PSI

19. Spring brakes do which of the following?
A. Come on when there is a leak in any part of the air brake system
B. Supply mechanical force to aid heavy vehicles in parking and emergency braking
C. Come on when there is a reduction of air pressure to 20 to 45 PSI
D. All of the above

20. When you are driving, ABS does which of the following?
A. Shortens your stopping distance
B. Increases your stopping distance
C. Helps you control your vehicle when braking hard
D. All of the above

21. You should not drive an air brake vehicle until the compressor has built up pressure to at least ________ for a truck.
A. 50 PSI
B. 100 PSI
C. 150 PSI
D. 75 PSI

22. If you are driving a truck, tractor or bus with ABS, you should do which of the following?
A. Watch your mirrors
B. Not change how you brake
C. Learn how to drive faster
D. Drive slowly at all times

23. Which of the following are two major techniques used for emergency stops?
A. Uncontrolled and controlled braking
B. Uncontrolled and stab braking
C. Controlled and stab braking
D. All of the above

24. Which of the following is the main instrument of the air brake mechanism?
A. Water vapor
B. Weather conditions
C. Atmospheric air
D. Speed and distance

25. Which of the following is not an inspection procedure for air brakes?

A. Inspect engine compartments.

B. Walk around and focus on checking the vital parts.

C. Inspect the disc linings.

D. Inspect the hose.

3 Doubles and Triples Test

1. How do you prevent rollover risk?
A. Drive slowly into curves.
B. Stop abruptly when there is danger in front.
C. Increase your speed to maintain balance.
D. Quickly change lanes when necessary.

2. When pulling doubles and triples, you should ____________.
A. Be vigilant.
B. Keep your eyes on the road.
C. Manage space.
D. All of the above.

3. How do you stop doubles and triples?
A. Abruptly
B. Slowly
C. Very fast
D. Don't stop them

4. What should you do if the weather is bad?
A. Wait for it to clear.
B. Quickly drive out of the weather.
C. Follow the shortest route out.
D. All of the above.

5. How should you park doubles and triples?
A. In a tight corner
B. In an area you can pull out from
C. Any way you like
D. Only in an enclosed parking area

6. Which of the following is not a step in inspecting the air brakes of doubles and triples?
A. Checking the flow of air
B. Inspecting the tractor control/protection valve
C. Inspecting the emergency brakes of the trailer
D. Checking the steering wheel

7. When you go to the rear of the rig and open the closed valve of the emergency line, what should you hear if you listen attentively?

A. Combustion

B. The engine sound

C. The sound of escaping air

D. A ringing bell

8. A brake leak signifies that _______________.

A. There is not enough engine oil.

B. There is an escape of air.

C. The tractor protection valve is malfunctioning.

D. The storage tank is empty.

9. What is the first thing you should do when inspecting the service brakes?

A. Keep the air pressure normal.

B. Close the parking brake.

C. Quickly drive the vehicle.

D. Release the trailer's brakes.

10. Which of the following is a step to coupling double trailers?

A. Ensure the safety of the second trailer at the back.

B. Integrate the converter dolly with the front trailer.

C. Integrate the converter dolly with the rear trailer.

D. All of the above.

11. Which of the following is used to couple the fifth wheel?

A. Fifth coupler

B. Wheel coupler

C. Converter dolly

D. Hand valve

12. Where should the converter dolly be in relation to the second trailer when coupling?

A. At the back

B. In front

C. At the side

D. Under

13. How do you couple the converter dolly with the front trailer?
A. Put the first semitrailer in front of the dolly tongue.
B. Connect the front trailer with the dolly.
C. Lock the pintle hook.
D. All of the above.

14. Which of the following is not a process for coupling the converter dolly with the rear trailer?
A. Lock up the converter dolly's air tank.
B. Lock up the valves behind the first trailer.
C. Lock up the fuel tank.
D. Lift the landing gear without restraints.

15. How do you power up the trailer's brakes?
A. Use the hand valve.
B. Apply the service brake.
C. Turn on the engine.
D. Push the air supply button.

16. When uncoupling twin or double trailers, you should do which of the following?
A. Set the rig firmly on clear ground.
B. Press the parking brake.
C. Lock up the second trailer wheels.
D. All of the above.

17. Uncoupling doubles involves disconnecting ___________.
A. The trolley valve and emergency brake
B. The air dolly and the electrical lines
C. The air dolly and the air compressor
D. The twin and the triple

18. You should let go of the _______ of the fifth wheel when uncoupling.
A. Latch
B. Door
C. Gate
D. Opening

19. Which of the following should you do to the wheel when uncoupling a converter dolly?
A. Combine
B. Brake
C. Chock
D. Wash

20. What can happen if the dolly is beneath the back trailer and you open the hook of the pintle?
A. There may be an explosion.
B. The dolly bar may fling up.
C. The converter dolly may stop working.
D. The pintle will explode.

21. Which of the following should you do when coupling triples?
A. Ensure the safety of the second trailer at the back.
B. Couple the converter dolly with the first trailer.
C. Couple the converter dolly with the back trailer.
D. All of the above.

22. Which of the following is not a step in inspecting doubles and triples?
A. Inspect the lower or bottom fifth wheel.
B. Sit down comfortably and let go of the arms while the safety lock is engaged.
C. Inspect the upper or top fifth wheel.
D. Do not place the glide plate on a frame.

23. What should you do with gaps between the top and bottom fifth wheels?
A. Open them.
B. Close them.
C. Fill them.
D. Reduce them.

24. Which of the following is not a step in inspecting the landing gear?
A. Check for damaged parts.
B. Maintain the crank handle.
C. Raise the gear halfway up.
D. Check for air leaks.

25. The valves at the back of the front trailer should be __________ when you are inspecting doubles and triples.

A. Closed

B. Open

C. Up

D. Down

4 Hazardous Materials (Hazmat) Test

1. Hazardous materials are ____________.
A. Objects that obstruct the movement of vehicles on the road
B. Items that inhibit drivers' sight on highways
C. Substances that are risky and may have adverse effects on safety, health and assets in the course of transportation
D. All of the above

2. According to Hazardous Materials Regulations, which of the following is true?
A. Every commercial vehicle transporting hazardous materials should have warning signs.
B. No commercial driver should transport hazardous materials on the highway.
C. Commercial drivers must have a college degree before they can transport hazardous materials.
D. All of the above.

3. Which of these is not a Class 1 hazardous material?
A. Fireworks
B. Dynamite
C. Acetone
D. Ammunition

4. Flammable solids are ____________.
A. Not too dangerous when exposed to fire
B. Not likely to combust into fire hazards
C. Easily removed from accident areas
D. None of the above

5. Even if you have a license, you are not allowed to carry which of the following?
A. Radioactive materials near where people are sitting
B. More than 500 pounds of hazardous materials
C. More than 100 pounds of a specific class of hazardous materials
D. All of the above

6. What's the intent of hazmat regulations?
A. To punish offenders
B. To monitor commercial drivers who disobey the rules
C. To keep the road clean
D. To reduce the danger to drivers, passengers and others

7. Which of the following statements is true?
A. Shippers should inform drivers about the risk involved in transporting hazardous materials.
B. Shippers should obtain the right shipping papers.
C. Shippers should provide information for emergency responses.
D. All of the above.

8. Which of the following is true about markings?
A. They help in the identification of cargo.
B. They need to be placed on portable tanks and cargoes.
C. They highlight the shipping name of every package on both sides.
D. All of the above

9. Which of the following should be your priority when marking the identification number?
A. Color
B. Weight
C. Design
D. Visibility

10. Whoever is loading hazardous material should stay ________ feet from the tank.
A. 15
B. 25
C. 30
D. 35

11. Transporting a leaking vehicle is ________.
A. Prohibited
B. An environmental hazard
C. A fire risk
D. All of the above

12. After successfully loading hazardous materials, you can ____________.
A. Open the package when necessary
B. Open the package to check if it is intact after 40 miles
C. Not open the package
D. Transfer the contents of the packages every hour

13. Bulk packaging that is not fixed on a vehicle is called a/an __________.
A. Portable tank
B. Cargo tank
C. Unfixed tank
D. Detachable tank

14. Which of the following statements is true about an MC306?
A. It is specifically meant for hazardous liquids.
B. It is used to transport hazardous rocks.
C. It supports hazardous gases.
D. It is a type of portable tank.

15. How many major stakeholders are there for the transportation of hazardous materials?
A. Two
B. Three
C. Four
D. Five

16. Which of the following is not the responsibility of a driver during the transportation of hazardous materials?
A. The transportation of products quickly and on time
B. Ensuring that the shipper has done due diligence
C. Transporting products that are leaking
D. Complying with every regulation in the transportation of hazardous materials

17. The communication rules necessitate which of the following?
A. Shippers should not highlight Class 1 hazardous materials in the shipping paper.
B. Phone contacts on the shipping papers should be available only in the daytime.
C. The hazmats shipping paper should be placed in a safe place where no one can easily see it.
D. None of the above.

18. How many classes of hazards are there?
A. Five
B. Three
C. Nine
D. Seven

19. Placards should not be less than ______ inches square.

A. 2

B. 10 3/4

C. 5 1/2

D. 3

20. Before you load hazardous materials in your vehicle, you should ensure which of the following?

A. The vehicle is warm.

B. The floor lining is tight and not made of metallic substances or of ferrous metal.

C. The engine is turned on.

D. The package is damp

21. Which of the following is not a precautionary way of handling hazmats?

A. Providing proper ventilation to materials that are susceptible to combustion

B. Loading corrosive materials one after the other and placing them upright

C. Placing nitric acid below other packages

D. Not bothering to inform the owner of a parking space before parking

22. Nitric acid should be placed _______.

A. Under other hazardous materials

B. Close to oxidizers

C. On top of other hazardous materials

D. Between gases and flammable solids

23. Ammonium picrate is an example of which class of hazardous materials?

A. Class 2

B. Class 3

C. Class 4

D. Class 5

24. A safe haven is _________.

A. A place where you can safely dispose of hazardous waste

B. An approved location where you can park your vehicle containing hazardous materials

C. A safe environment to which you can expose Class 4 and 5 hazardous materials

D. All of the above

25. A consignee is ___________.

A. The sender of a shipment delivery

B. The authenticator of a shipment's contents

C. The receiver of a shipment

D. The maker of a package

5 Combination Vehicles Test

1. Combination vehicles are different from other commercial vehicles because they ______.
A. Weigh less
B. Are bigger
C. Are faster
D. Have stronger steering wheels

2. When you steer a combination vehicle very fast, you can expect ____________.
A. To arrive at your destination very fast
B. A crack-the-whip tilting of your vehicle
C. The truck to gain more stability
D. All of the above

3. Which of the following factors is a danger?
A. Speed
B. Slow driving
C. Meticulous driving
D. All of the above

4. The different positioning of the front and rear wheels of your vehicle is called ________.
A. Off-tracking
B. Double wheeling
C. Multi wheeling
D. All of the above

5. To avoid getting in the way of pedestrians, how do you steer the front of a vehicle around a corner?
A. Very slowly
B. Very thinly
C. Very quickly
D. Very wide

6. The Johnson bar is also called ________.
A. The steering bar
B. The trailer hand valve
C. The brake tester
D. The English bar

7. Which of the following is true about foot brakes?
A. They transfer air to every part of the air brake system.
B. They reduce the possibility of skidding.
C. They are better than using a trailer hand valve when testing the hand brakes while driving.
D. All of the above.

8. If the air pressure is around ________, then the tractor protection valve will close up without manual application.
A. 20 to 45 PSI
B. 45 to 50 PSI
C. 50 to 60 PSI
D. 30 to 50 PSI

9. Which of the following is used to supply and shut off the air?
A. The hose couplers
B. The air lines
C. The glad hands
D. The air supply control

10. The service line is also called the ______ or _______ line.
A. Brake, combustion
B. Parking, emergency
C. Signal, control
D. Power, supply

11. Which of the following supplies air to the air tank?
A. Service air lines
B. Emergency air line
C. Control air line
D. Signal air line

12. The emergency line is often which color?
A. Green
B. Blue
C. Purple
D. Red

13. Which of the following is true of the hose couplers?
A. They couple the service and emergency lines in the truck/tractor with the trailer.
B. They are made of iron.
C. They help allow air escape.
D. All of the above.

14. Shut-off valves are also known as ________.
A. Close-ups
B. Cut-out cocks
C. Power shuts
D. Valve enclosures

15. What should you do when supervising the fifth wheel?
A. Check if the kingpin is damaged or out of shape.
B. Check the windshield.
C. Inspect the tires.
D. Turn on the engine while inspecting.

16. How high should your trailer be when coupling it?
A. Very low
B. Very high
C. Reasonably low
D. None of the above

17. Which of the following is not a step to the coupling of air lines to trailers?
A. Ensure the glad hands are intact and link the emergency line to the trailer's emergency glad hand.
B. Connect the service line and the emergency glad hand.
C. Ensure the protection of air lines so that they will not be crushed when the tractor is taking the back of the trailer.
D. Link the service air line to the trailer's service glad hand.

18. What should you do to provide air to the trailer when coupling combination vehicles?
A. Switch off the engine so you can hear the sound of the brakes.
B. Listen to the sound as you press and let go of the trailer brakes.
C. Inspect the pressure gauge to see if there is any air loss and ensure the air pressure is moderate.
D. All of the above.

19. When supervising the coupling, you should do which of the following?
A. Get a flashlight.
B. Ensure no gap exists between the lower and upper fifth wheels.
C. Make sure the lever for locks is in the lock position and the safety latch is over it.
D. All of the above.

20. There should be ____________ between the landing gear and end part of the tractor frame if the trailer is completely on the tractor.
A. No gap
B. Enough clearance
C. Six feet
D. None of the above

21. What kind of surface should you use in setting up the rig?
A. Light
B. Strong
C. Moist
D. Slippery

22. Why should you back up slowly in the second step of uncoupling trailers?
A. To gather momentum
B. To reduce the pressure on the fifth wheel
C. To service the tires
D. To test the brakes

23. Which of the following should you do when putting down the landing gear?
A. Take the landing gear down until it touches the ground.
B. Quickly turn the crank after making the landing gear touch the ground.
C. Lift the fifth wheel.
D. Take the landing gear up.

24. When uncoupling, what should you do after setting apart the air lines and the trailer?
A. Combine the air lines with the trailer.
B. Fix the air line glad hands with the dummy couplers.
C. Expose the lines.
D. All of the above.

25. Where should you stand when uncoupling the fifth wheel?

A. In front of the tractor wheels

B. In the middle of the tractor wheels

C. Far back from the tractor wheels

D. None of the above

6 Tanker License Test

1. You will need which of the following if your tanker is transporting gases or liquids fixed in a cargo?
A. Cargo license
B. Tanker license
C. CDL certificate
D. Certificate license

2. If your vehicle requires a __________ CDL, you will need a tanker license.
A. Class A
B. Class B
C. Class A or Class B
D. Class E

3. Which of the following is true about driving tankers?
A. Tanker vehicles are unlike other vehicles.
B. You need special knowledge and skills to drive a tanker vehicle.
C. The center of gravity is high.
D. All of the above.

4. A greater part of the weight of a tanker's contents is ________.
A. Low
B. Raised high
C. Diminished
D. Stable in the road

5. Where will a rollover most likely occur?
A. In curves
B. On straight roads
C. On dry roads
D. On asphalt roads

6. What should you do when loading and driving liquids?
A. Fill the tanker up to the brim.
B. Drive quickly to be on time.
C. Leave some space when filling the tanker.
D. Keep the vehicle low.

7. When there is a strong liquid wave, you should expect your tanker to tilt toward _______.

A. The left

B. The right

C. The direction of the wave

D. The ground

8. Liquid tanks need _________.

A. Partitioning

B. Contents loaded in large quantities

C. To be loaded to the brim

D. None of the above

9. Liquid surge _________.

A. Occurs sometimes.

B. Should be avoided

C. Is caused by the size of the tanker

D. Is the reserving of liquid until it is needed

10. Which of the following is false about baffles?

A. They should be used for transporting milk.

B. You should expect a very high movement of the tanker when transporting liquid and lots of surging of the liquid.

C. They are a tool for reducing the surging of liquid.

D. None of the above.

11. You need ___________ when loading liquids in tankers.

A. Height

B. Outage

C. Low quantity of liquid

D. All of the above

12. Liquids expand when they are _________.

A. Cool

B. Bottled

C. Warm

D. All of the above

13. Which of the following is not true about liquids?
A. There is a stipulated weight limit for liquids.
B. You should not fill up the tank if the liquid is dense.
C. The liquid may expand during transportation.
D. Some acids are not dense liquids.

14. How do you drive tankers safely?
A. Drive smoothly.
B. Control the liquid surge.
C. Drive slowly in curves.
D. All of the above.

15. To control liquid surge, you should ________.
A. Give the brakes consistent pressure.
B. Take it slow when you are stopping.
C. Keep a good following distance.
D. All of the above.

16. Which of the following helps you prevent a crash?
A. Stopping slowly
B. Applying stab braking
C. Having a good following distance
D. All of the above

17. Jackknifing may be a result of _________.
A. Driving slowly
B. Skidding
C. Low acceleration
D. Braking slowly

18. When should you inspect your vehicle?
A. After loading
B. After unloading
C. When driving
D. None of the above

19. Which of the following is true about leaks?
A. It is very risky to load and drive a vehicle with leaks.
B. It is illegal to load and drive a vehicle with leaks.
C. Authorities may mandate you to clean up any spill along the route you have been driving.
D. All of the above.

20. When should you fix leaking valves?
A. After loading
B. Before unloading
C. When driving the tanker
D. All of the above

21. Which of the following is not special purpose equipment?
A. The grounding wires
B. The bonding wires
C. The fire extinguisher
D. None of the above

22. Applying the brakes while driving a tanker on the steep downgrade of a mountain __________.
A. Makes driving easy
B. Is the most necessary step in braking
C. Supports the brakes
D. Prevents fading of the brakes

23. How do you communicate with other drivers when there is a hazard or obstacle?
A. Stop your vehicle.
B. Wave your hands.
C. Flash the emergency lights.
D. All of the above.

24. You should expect __________ to occur when driving a tanker on a wet road.
A. Hydroplaning
B. Combustion
C. Exhaustion
D. Acceleration

25. What should you do when driving down a hill?
A. Ride the brakes very hard.
B. Find an escape route.
C. Drive slowly in a gear you can control.
D. Accelerate.

7 Passenger Transport Test

1. Which of the following is true of passenger drivers?
A. It is not their duty to prevent riders from boarding the bus with unlabeled hazmats.
B. They should not load passengers.
C. They should not allow riders to carry common hazmats onto the bus.
D. None of the above.

2. When should a safety inspection be done?
A. After embarking on a journey
B. After loading passengers
C. Before loading and embarking on the journey
D. Only when hazardous materials or objects are about to be loaded

3. What should you do with the inspection report if there is a need for repairs?
A. Sign it and call emergency numbers immediately.
B. Refuse to sign it.
C. Sign it before taking the vehicle for repairs.
D. None of the above.

4. The reflectors and the lights __________.
A. Should always be inspected
B. Are not essential parts
C. Should be inspected once a month
D. Should not be inspected

5. Emergency exits should be _________ before a vehicle is driven.
A. Opened
B. Closed
C. Permanently sealed
D. None of the above

6. Passenger seats should be __________.
A. Loosened
B. Unfastened
C. Fastened
D. 20 inches long

7. The emergency lights should _________at night.
A. Rarely be turned on
B. Always be turned on
C. Always be switched off
D. Be turned on only if there is an emergency

8. Which of the following is essential?
A. Spare tires
B. Extra electrical fuses
C. Fire extinguisher
D. All of the above

9. Seat belts are ___________.
A. Essential components
B. Necessary only if the bus is filled with passengers
C. Not ever necessary
D. Required only by the driver

10. Which of the following is a procedure for loading a passenger vehicle?
A. Allowing passengers to place their baggage in the aisle
B. Informing riders of the existence of emergency exits
C. Allowing passengers to keep their baggage where they think is best
D. All of the above.

11. When there is an emergency, riders should ____________.
A. Not exit from the emergency window
B. Exit only through the closest door
C. Exit through the window or the closest door
D. Stay inside the bus

12. Which of the following statements is true about transporting hazardous materials?
A. Buses are allowed to transport all hazardous materials.
B. Without a hazmat license, you can transport hazardous materials.
C. It is not necessary to boldly mark hazardous materials.
D. Explosives should not be placed near passengers.

13. Small arms and ammunitions are tagged as __________.

A. ORM-D

B. SAM

C. AMR

D. Small AM

14. Solid poisons weighing _______ should not be transported.

A. Less than 100 pounds

B. Less than 50 pounds

C. More than 100 pounds

D. More than 20 pounds

15. Irritating materials are __________in passenger buses?

A. Allowed

B. Not allowed

C. Placed on the roof

D. Occasionally allowed

16. Which of the following shows passengers where they should and should not stand?

A. Seats

B. Standpoint

C. Standee line

D. Stay director

17. What do you do when arriving at bus stops?

A. Announce the destination.

B. State why you are about to stop.

C. Announce the time of departure before and/or after every stop.

D. All of the above.

18. Which of the following is an important step when you are driving a bus for hire?

A. Ensure no one steps into the bus until it is time to move.

B. Leave one seat unchecked.

C. Do not allow any hazardous materials.

D. All of the above.

19. Passengers must be made aware that _______ is prohibited on the bus.

A. Smoking

B. Drinking

C. Sexual intercourse

D. All of the above

20. When driving, you should always ________.
A. Caution offenders.
B. Correct those whose heads or arms are outside the windows.
C. Be vigilant.
D. All of the above.

21. How should you handle disruptive passengers?
A. Drop them off immediately.
B. Drop them off at the next bus stop.
C. Take them back to where they were picked up.
D. None of the above.

22. It is ________ to expect that other drivers will give you space after you have started pulling out.
A. Right
B. Wrong
C. Normal
D. Ideal

23. Which of the following is true?
A. You can always depend on your vehicle's speed to be the same as a car's speed.
B. Speed can lead to loss of life.
C. You should take curves fast.
D. None of the above.

24. When stopping at railroad crossings, you should apply the brake when you are about ________ feet from the crossing.
A. 3 to 7
B. 45 to 60
C. 15 to 50
D. 5 to 10

25. What should you do if a drawbridge does not have a traffic control attendant or green signal light?
A. Apply your brakes.
B. Slow down.
C. Make sure your vehicle is at least 50 feet from the drawbridge.
D. All of the above.

8 School Bus Endorsement Test

1. What should you do when there's a dispute in your vehicle?
A. Turn back to see who is causing trouble.
B. Take your hands off the wheel to settle the dispute.
C. Keep your gaze on the road.
D. All of the above.

2. When should you observe the school rules on discipline?
A. Before transit
B. During transit
C. After transit
D. All of the above

3. What is the best way to resolve conflict in your vehicle?
A. Switch parties from their seats.
B. Allow the parties to stay together after you talk to them.
C. Remove the arguing parties immediately.
D. Ask the parties in conflict to sit in the front seats.

4. When are accidents more likely to occur?
A. When passengers are on board
B. When getting off a bus or when boarding it
C. When children are sick
D. When the driver is meticulous

5. Which of the following do you use when approaching a designated bus stop?
A. The signal arm
B. The lights
C. The crossing control arm
D. All of the above

6. How many seconds before arriving at the school bus stop should you switch on the warning lights?
A. 5 to 10 seconds
B. 5 to 10 minutes
C. 40 to 50 seconds
D. One hour

7. How far ahead do you switch on the right-side indicator when you are about to pull over?

A. 3 to 5 seconds

B. 10 to 15 minutes

C. 20 to 55 seconds

D. One hour

8. Students should ______ to you when boarding the bus.

A. Run

B. Play

C. Walk

D. Wait

9. When stopping, the vehicle transmission should be set in _______.

A. Neutral or park

B. Left or right

C. Up or down

D. Neutral or left

10. Students should always ___________.

A. Wait for the school bus along the road.

B. Stay in their rooms to wait for the school bus.

C. Wait for the school bus to arrive at a specific bus stop.

D. Walk along the sidewalk until the school bus comes to pick them up.

11. Which of the following should you do at the bus stop?

A. Signal to the students to board.

B. Take a count of the students.

C. Ensure no one is missing.

D. All of the above.

12. How should the students enter the bus?

A. In twos

B. In threes

C. In a single-file line

D. All of the above

13. When should you embark on your journey?
A. Once everyone has entered the vehicle
B. Once it is time to depart
C. When everyone is seated
D. None of the above

14. Which of the following is a procedure for unloading at a roadway?
A. Stay on the bus and signal the students to cross.
B. Stay on the left side of the road.
C. Look only to the right.
D. None of the above.

15. How should students get off the bus?
A. However they like
B. Row by row
C. In fives
D. All of the above

16. What do you do with the alternating flashers when you are preparing to leave after loading?
A. Switch them on.
B. Switch them off.
C. Remove them.
D. None of the above.

17. The left zone of your school bus is considered ________.
A. A safe haven
B. The emergency spot
C. A danger zone
D. A restricted area

18. Which of the following is true about mirrors?
A. They should be properly adjusted.
B. They are useful in traffic.
C. You can use mirrors to check the welfare of students.
D. All of the above.

19. Which of the following is/are used to check incoming vehicles and activities going on outside the back of the vehicle?

A. The outside flat mirrors

B. The outside convex mirrors

C. The outside crossover mirrors

D. The overhead rearview mirror

20. When you properly adjust the outside flat mirrors, you'll be able to see _________.

A. The top of your vehicle

B. Beneath your vehicle

C. The sides of your vehicle

D. All of the above

21. Which of these mirrors will give you a view of the back tires where they make contact with the ground?

A. The outside flat mirrors

B. The outside convex mirrors

C. The outside crossover mirrors

D. The overhead rearview mirror

22. Which of the following is not true about the outside convex mirrors?

A. They will not give you the exact distance and size of objects.

B. You can use them to see what the students are doing outside.

C. They are exact in reflecting the size of people.

D. They give you a wide view of activities outside.

23. Which of the following gives you a good view of danger zones in front of and outside the bus?

A. The outside flat mirrors

B. The outside convex mirrors

C. The overhead rearview mirror

D. The outside crossover mirrors

24. When driving, which of these mirrors should you use?

A. The outside convex mirrors

B. The outside crossover mirrors

C. The outside flat mirrors

D. All of the above

25. Properly positioning __________ will give you a view of the back window that is on top of the mirror.

A. The overhead rearview mirror

B. The outside convex mirrors

C. The outside crossover mirrors

D. The outside flat mirrors

1 Answers for General Knowledge Test

(1) (C) It supplements or supports the brake effect the engine has on the vehicle.

When you apply the brakes on either the long or steep downgrade of a mountain, it supports the impact of the engine on braking. On the downgrade, you are expected to slowly reduce the speed of your vehicle.

(2) (D) To load every student first.

You should load every student on the bus before backing up to prevent crushing passengers or bystanders around the vehicle. After loading, you will have to find someone to help guide you in backing up. The role of a lookout is to inform you if you are about to hit a vehicle, an obstacle or a person. This person's duty does not include showing you how to back out.

(3) (D) Stop and drive inspection.

This is not a vehicle inspection procedure. The pre-trip inspection is carried out before driving. It helps you prevent a breakdown on the road. During the trip, inspection is carried out while driving. This includes checking the lights, the tires, the cargo, etc., at intervals. An after the trip inspection is carried out after driving.

(4) (B) Wait for enough of a gap.

Do not hurry to cross traffic. First, confirm if the space is long enough for your double and triple vehicle. Before you merge into another lane, confirm that no vehicle is next to you. These steps help avoid a collision.

(5) (B) Communicate with the emergency flashers to warn drivers behind you.

Use the brake lights' emergency flashers to warn other drivers when there is trouble ahead, when you have to take a tight turn, when you are slowing down on the road and when you are stopping.

(6) (D) Prevents your wheels from locking up.

An antilock brake system (ABS) helps prevent your wheels from locking up.

(7) (B) Move the gear to the right downshift.

Slow down and downshift to the right-hand side when approaching a curve. This helps you harness the curve for stability. With your gear in this position, you can also easily speed up after driving through the curve.

(8) (B) Hydroplane.

Water or slush on the surface of the road sometimes leads to hydroplaning. The tires lose traction because they do not have enough contact with the road. This may affect your ability to steer your vehicle or apply the brakes. To regain control, let go of the accelerator and press the clutch. Do not use the brakes to reduce your speed.

(9) (A) Safely stop runaway vehicles.

Escape ramps are made to reduce the speed of runaway vehicles on steep mountain downgrades. They are designed to prevent accidents and injuries. Without an escape ramp, there would be a loss of lives, equipment and cargo. Learn to identify the locations of these escape ramps.

(10) (C) The brakes will develop faults.

Too much heat on the brakes may make the brakes faulty. It becomes more difficult to control the brakes because you may have to increase the pressure you place on the handle before it responds to you. This means that your brakes are fading, and you will soon find it hard to stop your vehicle.

(11) (D) None of the above.

An ABS helps prevent your wheels from locking up. It is computerized and works with your normal brakes without reducing or increasing the effectiveness of your normal brakes. ABS ensures you maintain a reasonable speed. It is not a replacement for your normal brakes. Its sole function is to stop your wheels from locking up due to too much braking.

(12) (B) The safety of everyone when driving.

Safety is the priority of every driver. Before, during and after driving, you should perform a safety inspection to avoid crashes.

(13) (D) All of the above.

All of the answer options are inspection techniques when driving. They help you keep your vehicle safe and prevent unexpected situations that may lead to problems. They are done after the pre-trip inspection procedures and before the after-the-trip inspection procedures. Your safety on the road depends on these processes.

(14) (D) None of the above.

The reason for removing the keys from the ignition or starter before you start your pre-trip inspection is to prevent someone from moving the truck while you are still under it.

(15) (C) Accept the situation and arrive late.

Fast driving is dangerous. It may lead to the toppling of your vehicle, collision with other vehicles, frustration and road rage. A good driver accepts a delay and maintains the right speed and road rules. If there is a need to stop, take a break. What really matters is your safe arrival.

(16) (C) 50 miles.

Inspecting your cargo within the first 50 miles will help you ensure the safety and security of your equipment. Do not wait for the end of your trip. A part may need adjustment and repair along the way before it escalates to a critical danger. It is too soon to check at 10 to 25 miles, and 100 miles is too far.

(17) (A) The rear wheels will move to another path.
When your vehicle goes around a corner, there will be an off-track of the rear wheels. The rear wheels will take another path from the front wheels. This is also known as off-tracking or cheating. Expect longer trailers to have more cheating than shorter ones.

(18) (C) A four times increase in stopping distance.
An increase in speed greatly affects the stopping distance. Traveling twice as fast results in a four times increase in stopping distance. It is better to reduce the speed of your vehicle because the more difficult it is to stop your vehicle, the greater the risk of a crash.

(19) (A) Increase your following gap.
Tailgating is when another vehicle is dangerously close to your rear. When you are tailgated, you should increase your following gap. Gradually reduce your speed. Signal that you are slowing down and do not quickly swerve to another lane. Allow room in front of you because this will help the vehicle behind remain safe. Your brake lights and taillights should be switched off when you are being tailgated.

(20) (C) Speed.
An increase in speed greatly affects the stopping distance. The more difficult it is to stop your vehicle, the greater the risk of a crash.

(21) (A) Your vehicle has placards.
You should stop at railroad crossings if your vehicle is carrying chlorine, if there are loaded or empty cargo tanks for hazardous materials in the vehicle and if your vehicle is placarded. After you come to a stop, you can drive on when you are certain there is no oncoming train. Remember not to shift gears while crossing.

(22) (D) All of the above.
It is not easy to shift gears when using double clutching. Do not try to force the gear. The vehicle may already be in neutral. Do not allow it to stay in neutral for a long time. If it is difficult to switch to another gear, go to neutral, release the clutch and increase the speed of the engine to be in alignment with the road speed. Then try shifting the gear again.

(23) (D) None of the above.
When there is a fire in the vehicle, you should first turn off the engine immediately. Keep the hood closed. Use the louvers to shoot foam. If the cargo fire is in a box trailer and there are hazardous materials, close the doors to prevent oxygen from spreading the fire.

(24) (D) None of the above.
Too much or too little air pressure is not good for the tires.

(25) (B) Sleepiness.

When a vehicle lacks proper ventilation, this will make the driver groggy. Therefore, the vehicle's interior needs to have cool air. The windows should be open. If air conditioners are installed, turn them on.

(26) (C) Stop and take a nap.

Do not be in a hurry to keep up with the schedule. If you are tired or groggy, stop in a safe place and take a nap. Medication and/or coffee will not resolve fatigue.

(27) (A) Park your vehicle in a safe spot and resolve any problem.

When you notice a fault in your vehicle, do not drive any further. Instead, stop in a secure area and find out what is wrong. If there is a conflict between passengers, parking and resolving the conflict is the best thing to do. Never be in a hurry to arrive at your destination.

(28) (B) 15 feet.

The length you need to add to your school bus is 15 feet. You should know the size of a school bus's safety areas at highway crossings. When you are close to the signal, take a good look at the size of the space. Ascertain that there is sufficient room to clear the railroad tracks if you are required to stop.

(29) (C) You will be able to move forward when leaving.

Backing up should be avoided as much as possible. Look in your direction when parking. Use the mirrors to watch your side and rearview mirrors. There should be enough space to move forward when leaving the parking spot because sometimes pulling out can be difficult, especially if the space is too tight.

(30) (B) Not go too fast, and use a controllable gear.

When going down a hill, reduce the speed to what you can easily control. Go easy on the brakes to avoid overheating. Accelerating at a high speed down a hill will endanger passengers and cargo. The gear should not be equal to what it takes to go up the same hill; instead, you should lower the gear, or downshift.

(31) (D) All of the above.

A very slippery road can lead to accidents and injuries. The best thing to do is to stop driving and park in a safe place. If the road is not too slippery, you may decide to keep driving, but you should drive slowly and smoothly. If there is ice on the wiper blades, be wary of sliding.

(32) (A) If the road is wet or filled with snow.

Drivers can switch retarders on or off at different moments, including if the road is wet or filled with snow. Retarders are built in on some vehicles. They help reduce speed and are alternatives to brakes and will help prevent brake wear and tear. The four types of retarders are engine, electric, exhaust and hydraulic.

(33) (D) It should not be used in hot weather.

This is false. Before you start a trip, inspect the tires and be sure that a spare is available. Also check the condition of the spare tire. It should be inflated, not damaged, and must be the proper fit for the vehicle. Never go on a journey without a spare tire, regardless of the weather or temperature.

(34) (A) Heat.

When the weather is very hot, this is a sign you should check the tires every 100 miles or two hours. High temperatures often increase air pressure. If you discover that a tire is becoming too hot, stop and wait for it to cool off to prevent a tire explosion. Do not try to reduce the air in the tire.

(35) (C) The vehicle will be deemed out of service.

According to federal and state regulations, drivers are mandated to inspect their vehicles before, during and after a trip. When federal and state inspectors discover that a vehicle is unsafe, it will be deemed out of service until the vehicle is repaired. This is done to ensure the safety of the driver, the passengers and the packages being transported.

(36) (B) A faulty ABS means a truck will depend solely on normal brakes.

This is true. An ABS helps prevent your wheels from locking up. Therefore, a faulty ABS means the truck will depend solely on normal brakes

(37) (A) Wheel brake.

This is not a type of brake.

(38) (D) All of the above.

Three rules govern the use of turn signals—signal early, signal continuously and cancel the signal. Before you take a turn, set a noticeable signal and do not turn it off until you have completed the turn. Communicating your intentions through signals will help ensure safety and prevent a crash.

(39) (D) All of the above.

Before you start driving, ensure there is sufficient water and antifreeze in the engine cooling system. When there is a need to add coolant to the system, turn off the engine, wait for it to cool, turn off the radiator slowly and step back to wait for the pressure to be released from the cooling system.

(40) (C) Are not safe.

During inspection, check for damaged rims. Wheels or rims should not be welded. Instead, they should be replaced immediately, for this is the best way to prevent accidents. The rims and wheels are two of the most used parts of a vehicle. Therefore, they need to be optimally functional.

(41) (D) The road becomes very slippery.
When a road is very slippery, it is the worst time to drive. To ensure the safety of your vehicle and that of everyone/everything on board, stop and find a safer route. If there is no safer route, reduce the speed of your vehicle and do not brake in a curve.

(42) (C) Three.
Single-air systems require pressure to go from 50 to 90 PSI in three minutes at an idle speed. However, if it is a dual-air system, the pressure goes from 85 to 100 PSI in 45 seconds. A larger air tank takes a longer time to build up air pressure.

(43) (D) All of the above.
When on a trip, you need to check your vehicle's condition. You should not wait until you get to your final destination. You need to be vigilant with your eyes, nose and ears. Listen for danger. Smell for a fuel leak. Use your eyes to see if gauges and components are working.

(44) (B) The parking brake should be off.
This is not true. You should make sure the parking brake is on when you inspect the compartment engine. The wheels should be chocked, and the hood raised. Check the oil level, coolant level, power steering fluid level, hydraulic level and fluid level in the battery. Inspect for leaks in the engine compartment.

(45) (B) Dual.
A lot of health-duty vehicles are operated with dual air brake systems to ensure safety. This kind of air brake comprises two different systems, which share a single brake control. Each of the systems has its own separate hoses, lines, air tanks, etc. If there is a trailer, the dual system gives it an air supply.

(46) (D) You should not move the vehicle into oncoming traffic.
When you are ready to turn, check traffic in every direction, place both hands on the steering wheel, keep checking your mirrors and remain in position if there is oncoming traffic. Do not change gears while turning.

(47) (D) None of the above.
All of these answers are false. Water or slush on the surface of the road sometimes leads to hydroplaning. The tires lose traction because they do not have enough contact with the road. This may affect your ability to steer your vehicle or apply the brakes.

(48) (A) Set the front wheels facing forward.
If you need to stop before taking a turn, the stop should be smooth, without skidding, and it should be completely done before the stop sign, stop line or crosswalk. When behind a vehicle, stop where you can see that vehicle's back tires. Do not allow your vehicle to roll, and always set the front wheels facing forward.

(49) (A) Get everyone out of the bus immediately.

To prevent an escalation of the problem and a collision with other vehicles, get everyone out of the bus and off the tracks immediately. Keep everyone at a site that is far away from the bus and the tracks. This will ensure the safety of every passenger.

(50) (D) All of the above.

Your capability to handle a heavy vehicle will be tested based on how you drive the vehicle to the front, back, left and right; to specific angles and toward a particular destination. You should expect to be tested on some markings with cones, traffic lanes, barriers and other objects.

2 Answers for Air Brakes Test

(1) (D) All of the above.

Air brake systems are often found in trucks, heavy vehicles and commercial buses. It requires a lot of effort to operate an air brake system. The weights and sizes of these vehicles need the mechanism that air brakes provide. One of these mechanisms is the use of compressed air pressure to power the brakes. Heavy-vehicle drivers need air brakes to efficiently and safely stop their vehicles.

(2) (D) All of the above.

Human effort and concentration alone cannot operate heavy vehicles. You also need mechanical efforts to power these vehicles. For air brakes, the system uses compressed air to supply greater force. The generated force creates friction against the brakes and the tires through the brake pads.

(3) (C) Parking brake, service brake and emergency brake.

Air brakes are sometimes called tripartite brakes because they have three parts in one. These parts perform complementary functions. They are the parking brake, the service brake and the emergency brake. The parking brake is used when attempting to pull over or park. The service brake is what you use in your usual driving routine. And the emergency brake is used when there is a fault in the brake system.

(4) (D) All of the above.

The air brake system functions this way: The exhaust channel closes when there is pressure on the brake pedal. A closure results in the opening and passage of air along the air channel and the brake chamber, and the used air is released.

(5) (A) Place the drum line brakes on the wheels.

Whenever you intend to stop a vehicle, you should place the drum line brakes on the wheels. Friction occurs when brake pads impact the rotation. This friction results in the stopping of the vehicle. Expect that whenever you apply the brake pedals, the triple valve's outlet will close up, while the inner valve opens to let the compressed air in the air reservoir travel around the brake lines.

(6) (D) The air compressor.

The air compressor is the main component of an air brake system that uses a belt drive. Its purpose is to compress air in the environment to a particular pressure and take it to the air reservoir or storage tank. Without the air compressor, it is not possible to get air from the atmosphere into the storage tank.

(7) (B) The atmospheric air enters the storage tank.

The storage tank serves as a reservoir for atmospheric air that has been compressed under high pressure. It is a key component of the air brake system because it sustains the movement of a vehicle on a trip when there is a lack of atmospheric air needed for the operation of the vehicle.

(8) (A) Also known as the reservoir.

The storage tank makes it possible for the brakes to be applied as often as necessary. It serves as a reservoir for atmospheric air that has been compressed under high pressure. It is vital because it sustains the movement of a vehicle when there is a lack of atmospheric air for the operation of the vehicle.

(9) (B) They are used for stopping the vehicle in transition.

The driver uses the brake pedal to stop the vehicle. This is an essential part of the air brake system because a moving vehicle needs the brake pedal in order to stop. It requires human effort to operate. Therefore, it is manually operated in heavy vehicles.

(10) (B) The safety valve.

The storage tank needs to be secured, and that is what the safety valve is meant to do. You can find it attached to the air storage or reservoir tank. The safety valve makes sure the tank does not burst when there is excessive pressure from the constant air supplied by the compressor.

(11) (C) Carries out the action after brake application.

When you apply the brake, the brake drum carries out the action. It is present in all brake systems. You can see it attached to the tires of vehicles. The brakes require the brake drum to stop the vehicle.

(12) (B) When driving on a slippery road.

This component is properly labeled as normal and slippery to enable better control of it. The pressure that travels to the front brake will be cut in half if you switch to slippery mode. A limiting valve inhibits the front wheel from slipping when driving on slippery roads. If you put the control on normal, it will be able to stop normally.

(13) (D) All of the above.

The air filter and air dryer are complementary. They work on the atmospheric air before it is passed on to the air compressor. The air filter separates dust particles from the atmospheric air. The air dryer removes moisture from the air. Every efficient brake system needs clean and dry air, or else dirt particles will accumulate.

(14) (A) Two.

There are two types of air tank draining valves—the manual draining system and the automatic draining system. If it is a manual draining system, you will have to drain it by pulling a cable or turning a dial in a quarter-turn direction. The automatic draining system often has inbuilt electric heaters.

(15) (D) None of the above.

None of these answers are correct. The S-cam operates when you apply the brake pedal. Applying the brake pedal allows air into the brake chamber, and the pressure from the air makes the rod protrude. This affects the slack adjuster, which twists the shaft of the brake cam and rolls the S-cam.

However, when you let go of the brake pedals, the S-cam rolls back and brings the brake shoes back from the brake drum.

(16) (A) The wedge brake.

The wedge brake works by having the brake chamber push the wedge against the rear brake shoes. Within the brake drums, it pushes them in two. Some wedge brakes have just one brake chamber, while others have two. These brakes operate both manually and automatically.

(17) (A) The air tank.

With a supply pressure gauge, you can know the exact amount of pressure in the air tank. Air brake systems are designed to link a pressure gauge to the air tank. If it is a double-gauge system, then expect a gauge attached to each part of the system. It may even have just one gauge that comes with dual needles.

(18) (D) 80 to 85 PSI.

Most big buses give off a warning when air pressure is at 80 to 85 PSI. A low air pressure signal tells you when your pressure supply is almost below 60 PSI or when one-half of your air compressor governor is running down on pressure. You will see a red light and sometimes hear a buzzer. This warning means you have to attend to the situation.

(19) (D) All of the above.

Spring brakes supply mechanical force to help heavy vehicles park and stop in an emergency. They come on when there is a leak in any part of the air brake system or when there is a reduction of air pressure to 20 to 45 PSI. They are used for preventive purposes.

(20) (C) Helps you control your vehicle when braking hard.

An ABS helps prevent your wheels from locking up when you brake hard. It is computerized and works with your normal brakes without reducing or increasing their effectiveness.

(21) (B) 100 PSI.

Dual air brake systems are often used for safety purposes. Although the air brake system is double, it uses just one set of controls. Each comes with separate air brake parts, such as hoses, brake lines, air tanks, etc. One takes over the front axle while the other operates the back axle, and both provide air to the vehicle. The minimum pressure should be 100 PSI for a truck.

(22) (B) Not change how you brake.

Do not change how you brake when driving a truck, tractor or bus with ABS. The only exception to this rule is when you are handling a straight truck or combination vehicle with ABS on both axles. If an emergency stop is required, you can use the brakes without restraint.

(23) (C) Controlled and stab braking.

You can apply either controlled or stab braking techniques in an emergency stop. If a driver ahead is pulling over, you can simply step on the brakes if there is enough distance between your vehicle and theirs. Stop in a straight line using the controlled or stab braking technique.

(24) (C) Atmospheric air.
Air brakes are best for heavy vehicles because they use atmospheric air to supply pressure. Air brakes are classified as high-powered brake systems. They are often found in trucks, heavy vehicles and commercial buses. Using their legs, drivers apply great effort in operating this system. An air brake system will not work if accumulated pressure is not applied to the brake component.

(25) (B) Walk around and focus on checking the vital parts.
Option B is not an inspection procedure for air brakes. The whole system (not just the vital parts) needs to be inspected in an air brake. No part is too small to overlook. Every part plays a useful role that affects the whole system. So, walk around the vehicle and take note of the smallest details. If you notice any of the components are faulty, repair or replace them as soon as possible.

3 Answers for Doubles and Triples Test

(1) (A) Drive slowly into curves.

To avoid a rollover, drive slowly when taking a curve and ensure the cargo is not far from the ground. Do not change lanes abruptly. Fast driving causes rollovers. Steering gently helps prevent a crack-the-whip tilting of your trailer. Be gentle when pulling over and never make a sudden move.

(2) (D) All of the above.

Your mind and physical body need to be alert and always expecting the worst. Do not let down your guard. Keep your eyes on the road. Watch out for speed bumps, traffic, curves and other unforeseen factors that may pose an obstacle to the smooth movement of the double and triple trailers. Also, maintain a good distance between you and other vehicles.

(3) (B) Slowly.

It takes a few seconds before doubles and triples can be fully stopped. Therefore, do not get too close to other vehicles. Maintain a good distance or space. Do not hurry to cross traffic. First, confirm if the space is long enough for your double or triple vehicle. Unlike other kinds of commercial vehicles, doubles and triples occupy more space on the road because they are so long.

(4) (A) Wait for it to clear.

If the weather conditions are not ideal for driving, you should either wait for conditions to clear up or be extra careful on the road. Slippery roads and fog often lead to accidents. A double and triple vehicle requires greater skills in these adverse conditions.

(5) (B) In an area you can pull out from.

Where and how you park your vehicle is very important. You should park your vehicle in an area that you can easily pull out from. Assess the parking space to see if it is a fit for your vehicle. Careless parking may lead to difficulty when pulling out.

(6) (D) Checking the steering wheel.

Option D is incorrect. When inspecting the air brakes of doubles and triples, you should check the flow of air, inspect the tractor control/protection valve, inspect the trailer's emergency brakes and inspect the trailer's service brakes. These steps are to be combined with the steps for inspecting combination vehicles.

(7) (C) The sound of escaping air.

When you go to the rig at the rear and open the closed valve of the emergency line, you should hear the sound of escaping air if you listen attentively. This shows that you have supplied power to the system. After this, you can shut the valve of the emergency line and free up the service line.

(8) (C) The tractor protection valve is malfunctioning.

When there is a malfunction in the tractor protection valve, there could be a brake leak in which the air transported from the tractor is released. This makes the emergency brakes pop out. If care is not taken, there might be a loss of control of the vehicle.

(9) (A) Keep the air pressure normal.
When inspecting the trailer's service brakes, first ensure that the air pressure is normal, then open the parking brakes. Gently drive the vehicle to the front, then press the trailer's brakes using the trolley valve or hand control. You will observe that the brakes begin to come up.

(10) (D) All of the above.
When coupling doubles, ensure the safety of the rear trailer. Integrate the converter dolly with the front trailer, and integrate the converter dolly with the rear trailer.

(11) (C) Converter dolly.
A converter dolly is a tool for coupling one or two axles together, including a fifth wheel. You can couple the semitrailer to the back of a tractor-trailer. The second trailer should be behind the converter dolly. The second trailer should be behind the converter dolly.

(12) (B) In front.
When coupling, the converter dolly should be in front of the second trailer. To do this, open the air tank to let go of the dolly brakes. Use the parking brake to do this if there are spring brakes. Use your hand to wheel the dolly into alignment with the position of the kingpin. You can lift the converter dolly through the front semitrailer and the tractor.

(13) (D) All of the above.
When coupling the converter dolly with the front trailer, put the first semitrailer in front of the dolly tongue. Connect the front trailer with the dolly. Lock the pintle hook and safely set the converter gear into a lifted position.

(14) (C) Lock up the fuel tank.
Option C is false. When coupling the converter dolly with the back trailer, ensure that you have locked the trailer brakes and chocked the wheels. Then set up the height of the trailer. The converter dolly should be set beneath the back trailer. Safely set the landing gear in a lifted position. Lock up the converter dolly's air tank. Lock up the valves behind the first trailer.

(15) (D) Push the air supply button.
Press the air supply button to charge up the trailer's brakes. Then open the locked-up emergency line to check for air at the back of the second trailer. The absence of air means the brakes will not work because there is a mechanical fault somewhere.

(16) (D) All of the above.

When uncoupling twin or double trailers, set the rig firmly on clear ground and within a strategy sequence. Press the parking brake. If your vehicle does not have spring brakes, lock up the second trailer wheels. Release the landing gear of the rear semitrailer. Lock up the shut-off behind the front trailer. Remove all connections between the air dolly and the electrical lines.

(17) (B) The air dolly and the electrical lines.

When uncoupling twin or double trailers, you should remove all connections between the air dolly and the electrical lines.

(18) (A) Latch.

When uncoupling, let go of the dolly brake and the latch of the fifth wheel after you have removed all connections between the air dolly and the electrical lines. Then gently move the tractor. Note that the front trailer will be pulled first and the dolly out of the back of the semitrailer.

(19) (C) Chock.

When uncoupling a converter dolly, you need to press the spring brakes of the converter or chock the wheels to prevent the vehicle from moving. Remember to bring down the landing gear of the dolly. Remove the safety chains. Unlatch the pintle hook, which is on the front semitrailer. Then gently drive away from the dolly.

(20) (B) The dolly bar may fling up.

After chocking the wheels, you should unlatch the pintle hook, which is on the front semitrailer. If the dolly is beneath the back trailer, do not open the hook of the pintle. Otherwise, expect the dolly bar to fling up. This may result in injury or make it hard to recouple.

(21) (D) All of the above.

When coupling triples, use the same process for joining tractors to semitrailers. Ensure the safety of the second trailer at the back. Integrate the converter dolly with the front trailer and integrate the converter dolly with the back trailer.

(22) (D) Do not place the glide plate on a frame.

Option D is incorrect. When inspecting doubles and triples, inspect the lower or bottom fifth wheel. Ensure the glide plate is safely placed on the frame.

(23) (B) Close them.

When inspecting doubles and triples, you should close up gaps that exist between the top and bottom fifth wheels.

(24) (C) Raise the gear halfway up.

This is not a step in inspecting the landing gear. When inspecting the landing gear, keep the gear totally lifted up. Check for missing or damaged parts that need replacement or repair. Do not wait until later—do this immediately. Remember to securely maintain the crank handle. Watch out for air leaks, and quickly fix the leaking component.

(25) (B) Open.

When inspecting doubles and triples, the valves at the back of the front trailer should be open, while the valves at the back of the last trailer should be closed. The valve stem on the converter dolly's air tank should be closed. Ensure all air lines are well fixed and there is a secure connection of the glad hands.

4 Answers for Hazardous Materials (Hazmat) Test

(1) (C) Substances that are risky and may have adverse effects on safety, health and assets in the course of transportation.

Examples of hazardous materials are explosives, solids, gases, flammable liquids and other harmful materials.

(2) (A) Every commercial vehicle transporting hazardous materials should have warning signs.

This is true. The regulations are instructions for every commercial driver transporting hazardous materials. They mandate the use of warning signs, which are known as placards.

(3) (C) Acetone.

Acetone is not a Class 1 hazardous material. Fireworks, dynamite and ammunition are examples. Explosives are potential dangers to the safety of everyone in the vicinity. Do not allow smoke or fire to come close to your vehicle if there are explosive materials inside. Warn others to keep away if there is an accident.

(4) (C) Easily removed from accident areas.

Matches are examples of flammable solids. They fall under the Class 4 category of hazardous materials.

(5) (D) All of the above.

There are regulations regarding the number of hazardous materials a shipper is permitted to transport. You are not allowed to carry radioactive materials near where people are sitting, more than 500 pounds of hazardous materials, more than 100 pounds of a specific class of hazardous materials and explosives near where people are sitting, unless the cargo consists of small ammunition.

(6) (D) To reduce the danger to drivers, passengers and others.

Hazmat regulations exist for everyone's good. Their intent is to reduce the danger of hazmats, to inform drivers of the risks involved and to ensure the safety of equipment and the drivers who are directly in contact with these hazardous materials.

(7) (D) All of the above.

Shippers need to inform drivers about the risk involved in transporting hazardous materials. Hazmat regulations instruct shippers to place a warning on every package the drivers are transporting, obtain the right shipping papers, provide information for emergency responses and place placards.

(8) (D) All of the above.

Markings help in identification. They should be placed on portable tanks and cargo tanks. They must highlight the shipping name of every content on both sides. The size of the shipping name should not be less than two inches in height if the capacity of the portable tank is more than 1,000 gallons and one inch tall if the capacity is less than 1,000 gallons.

(9) (D) Visibility.

Markings need to be visible to everyone. Color and design are not the main factors to be considered when making markings. Your priority should be how legible the inscriptions are. Ensure the identification number is on both sides and on the end of a bulk package that has a capacity of more than 1,000 gallons.

(10) (B) 25.

Loading a hazardous product requires great care to avoid accidents. While loading, be circumspect when handling containers that are made up of hazmats. Avoid using materials that may puncture or damage the package. The person loading the package should stay within 25 feet around the tank.

(11) (D) All of the above.

Transporting leaking vehicles is against the hazmat regulations. Leaking vehicles are a potential danger for everyone. Fix any leaks before you begin your journey. The leaking of fuel or any combustible material can result in fire.

(12) (C) Not open the package.

Once you have successfully loaded and begun transit, do not open or transfer any package until you reach the destination. Extreme heat from the sun, smoke and other factors in the environment can set hazardous materials on fire. Loading and unloading should be done when the vehicle is parked.

(13) (A) Portable tank.

Bulk packaging that is not fixed on a vehicle is called a portable tank. When loading and unloading the packages, the portable tank is not on the vehicle. After the loading and unloading process, the portable tank is placed on the vehicle for movement. Every portable tank should have the owner or lessee's name.

(14) (A) It is specifically meant for hazardous liquids.

There are different types of cargo tanks. The most popular one is the MC306, which is meant for hazardous liquids, and the MC331, which supports gases.

(15) (B) Three.

There are three key stakeholders for the transportation of hazardous materials. They are the shipper, the carrier and the driver. The shipper is in charge of obtaining shipping papers and transporting goods. The carrier is in charge of proper documentation and accountability to agencies. The driver is mainly in charge of driving packages from one point to another.

(16) (C) Transporting products that are leaking.

This is not the driver's responsibility. The driver is responsible for ensuring that the shipper has done due diligence in identifying, marking and labeling hazardous materials; refusing to drive products that are leaking; and placing necessary placards. Drivers are also in charge of transporting products quickly and on time and safely keeping the documents of endorsement and other vital credentials.

(17) (D) None of the above.

The communication rules necessitate that every shipper highlight the types of hazardous materials that are in their shipment. They should provide phone numbers that are available 24/7. The hazmat shipping papers should be placed where they can be easily found in an emergency in which the driver and carrier are unresponsive.

(18) (C) Nine.

There are nine classes of hazards—explosives, flammable solids, oxidizers, corrosive liquids, flammable gases, flammable liquids, poisons, radioactive substances and organic peroxides. Hazmats pose a critical risk to drivers, passengers and citizens. Therefore, the government at all levels regulates the acquisition and use of these hazardous materials. They have procedures for loading and unloading because of the materials' volatile nature.

(19) (B) 10 3/4.

Placards are used to signal other drivers of the presence of hazardous materials. A vehicle that has been placarded should not have less than four identical placards. These are placed in the front, back and sides of the truck. The inscriptions on placards should be readable from four directions. They should be not less than 10 3/4 inches square and need to be diamond shaped.

(20) (B) The floor lining is tight and not made of metallic substances or of ferrous metal.

This is one way of ensuring safety while driving hazmats. Other ways of preventing damage or explosion of hazardous materials include switching off your engine before attempting to load or unload, cutting off heat sources and being extremely careful when handling explosives. Do not throw or drop the containers. Also, if a package is damp or stained with oil, you should not transport it. Label radioactive materials as Yellow III.

(21) (D) Not bothering to inform the owner of a parking space before parking.

Before you park your hazmat vehicle, you should always first inform the owner of the parking space. The owner will inform you whether the environmental conditions can accommodate hazardous materials. Remember to switch off your engine before attempting to load or unload hazardous materials.

(22) (A) Under other hazardous materials.

Nitric acid is a hazardous material. It should be placed under other materials. When loading corrosive liquids, which are hazardous materials, do not load them close to explosives, flammable solids, oxidizers and poisonous gases. This is to prevent potential dangers, such as explosions and fire.

(23) (C) Class 4.

Class 4 hazardous materials are flammable gases that are combustible when wet.

(24) (B) An approved location where you can park your vehicle containing hazardous materials.

You can always park your vehicle in approved locations, also known as safe havens. Local authorities are in charge of designating areas that should be used as safe havens. You may want to find a safe haven around your route and vicinity so that you can stop when necessary.

(25) (C) The receiver of a shipment.

A consignee is the receiver of a shipment. It may be a business or a person who is licensed to store hazardous materials.

5 Answers for Combination Vehicles Test

(1) (B) Are bigger.

Combination vehicles differ from other commercial vehicles in certain ways. They weigh more and are bigger, and they demand greater driving dexterity. Driving a normal bus or single vehicle is not the same as driving combination vehicles. Combination vehicles include straight trucks, trailers, doubles, triples and tractor-trailers.

(2) (B) A crack-the-whip tilting of your vehicle.

Steering fast will cause a crack-the-whip tilting of your trailer. Be gentle when pulling over and never make a sudden move. To keep your vehicle stable, drive slowly when taking a curve and ensure the cargo is not far from the ground. Do not change lanes abruptly. Fast driving causes rollovers.

(3) (A) Speed.

Take it slow with your trailer. Speed leads to fatal accidents. It takes longer to stop empty combination vehicles than it takes to stop loaded ones.

(4) (A) Off-tracking.

The front and rear wheels are often in different positions when you are going around a corner. This is called cheating or off-tracking. Expect longer trailers to have more cheating than shorter ones.

(5) (D) Very wide.

When the vehicle has gone off track and you want to come back to alignment, steer the front very wide around the corner so that the back end does not hit the curb, pedestrians or anything else in the way. Nonetheless, the rear of your vehicle should not be far from the curb. Other vehicles will not be able to pass you on the right.

(6) (B) The trailer hand valve.

Some people prefer to call the Johnson bar the trailer hand valve or the trolley valve.

(7) (D) All of the above.

The trailer hand valve is used to test the brakes and should not be engaged while driving because it may later skid. It is better to use the foot brakes in driving because they transfer air to every part of the air brake system and reduce the possibility of skidding.

(8) (A) 20 to 45 PSI.

The tractor protection control valve maintains the air in the air brake system if there is a leak in the trailer or if it breaks. If the air pressure is around 20 to 45 PSI, then the tractor protection valve will close up without manual application. This closure prevents air from escaping the air brake system and moves it through the emergency line.

(9) (D) The air supply control.

The air supply control is used to supply and shut off the air. When you push the control, there is an air supply and when you pull it out, the air supply is shut off. You can identify this component by its eight-sided red knob found in modern vehicles. It controls the air supply system.

(10) (C) Signal, control.

The service air line is also called a signal or control line. It transfers air around the air brake system. The service line pressure is based on the effort you exert on the brake or hand valve. The service line is linked to the relay valves, which facilitate the rapid application of trailer brakes.

(11) (B) Emergency air line.

The emergency air line is also called a supply line. It has two basic functions. The first purpose is to supply air to the air tank. The second is to control the emergency brakes.

(12) (D) Red.

Unlike service lines, which are most times blue, emergency or supply lines are often red.

(13) (A) They couple the service and emergency lines in the truck/tractor with the trailer.

The glad hands, or hose couplers, are tools that help couple the service and emergency lines in the truck/tractor with the trailer. The couplers come with rubber seals that help prevent air from escaping. Always ensure the rubber is clean before connecting the trailer with the truck.

(14) (B) Cut-out cocks.

Shut-off valves, or cut-out cocks, are for the supply or emergency line and the service air line, which are located behind towing trailers. These valves allow the closure of the air lines when there is no towing trailer. This is to prevent air leaks in the vehicle.

(15) (A) Check if the kingpin is damaged or out of shape.

When the fifth wheel is sliding, lock it up. Inspect the security and completion of your tractor. Check to see if the fifth wheel is well greased to avoid frictional steering problems. Inspect the fifth wheel position and set it in the proper coupling position. Lock up the fifth wheel if it is sliding. Then check if the kingpin is damaged or out of shape.

(16) (C) Reasonably low.

The trailer should be reasonably low when coupling so that the tractor can bring the trailer up when it is backed below the trailer. A very low trailer may make the tractor hit the trailer and destroy the nose. A very high trailer may make coupling difficult.

(17) (B) Connect the service line and the emergency glad hand.

This is not a step in the coupling of air lines to trailers. While coupling the air lines to the trailer, you should make sure the glad hands are intact and link the emergency line to the trailer's emergency glad hand. Connect the service air line to the trailer's service glad hand.

(18) (D) All of the above.

Among other procedures, you can provide air to the trailer by switching off the engine. Listen to the sound as you press and let go of the trailer brakes. Then inspect the pressure gauge to see if there is any air loss and ensure the air pressure is moderate.

(19) (D) All of the above.

When supervising the coupling, you should get a flashlight; ensure no gap exists between the lower and upper fifth wheels; check the back of the fifth wheel; ensure there is a closure of the fifth wheel jaw along the kingpin shank; lock the lever and fix any aspect of coupling before you start driving.

(20) (B) Enough clearance.

When applying the landing gear, ensure that there is enough clearance or distance between the landing gear and the end part of the tractor frame if the trailer is completely on the tractor. This is to make sure that the landing gear does not hit the tractor when there is a sharp turn.

(21) (B) Strong.

When setting the rig in the right position, first check the parking area and ensure its surface is strong enough to take the weight and size of the trailer. Remember that not all surfaces are good supports for the weight of trailers. Establish an alignment between the tractor and the trailer.

(22) (B) To reduce the pressure on the fifth wheel.

When reducing the pressure against the locking jaw, first switch off the air supply to lock the brakes, then back up slowly to reduce the pressure on the fifth wheel. Turn on the trailer's parking brake when the tractor is impacting the kingpin.

(23) (A) Take the landing gear down until it touches the ground.

Bring down the landing gear until it touches the ground. After the trailer is loaded, turn the crank slowly to cut off the tractor's extra weight. The fifth wheel should be left down so that it will not be difficult to unlatch it.

(24) (B) Fix the air line glad hands with the dummy couplers.

When disconnecting electrical cables from the air lines, first set apart the air lines and the trailer, then fix the air line glad hands with the dummy couplers, which are at the rear of the vehicle. Hang the electrical cable to avoid moisture. Support the lines to avoid incidents while on the road.

(25) (C) Far back from the tractor wheels.

When uncoupling the fifth wheel, lift the release handle lock and place it in the open position. Stand far back from the tractor wheels to prevent injury that may occur if the vehicle moves.

6 Answers for Tanker License Test

(1) (B) Tanker license.
You will need a tanker license if your tanker is transporting either gases or liquids that are in a fixed cargo.

(2) (C) Class A or Class B.
Every vehicle that requires a Class A or Class B CDL needs a tanker license. You will need a tanker license if your tanker is transporting either gases or liquids that are in a fixed cargo. The cargo tank capacity may be 119 gallons or more, or it can be a portable tank that has a capacity of 1,000 gallons or more.

(3) (D) All of the above.
You need special knowledge and skills to drive a tanker vehicle. This is because there is great liquid turbulence and an accentuated center of gravity involved. Therefore, you should not apply the same driving method to tankers that you use when driving other vehicles.

(4) (B) Raised high.
A greater center of gravity means that a greater part of the weight of the contents is raised high and almost off the road. Therefore, the top of the vehicle becomes weighted and is likely to tip over.

(5) (A) In curves.
Rollovers are most likely to occur in curves. Therefore, curves should be taken slowly.

(6) (C) Leave some space when filling the tanker.
When loading tankers with liquid contents, do not fill the tanker up to the brim. Leave some space in the expectation of a liquid surge.

(7) (C) The direction of the wave.
If a strong wave hits the rear of the tanker, expect the tanker to tilt toward the direction of the wave. If the vehicle is parked on slippery ground, the wave can move the vehicle in its direction.

(8) (A) Partitioning.
A bulkhead is used to partition liquid tanks into smaller units to avoid an excess load on one part of the vehicle.

(9) (A) Occurs sometimes.
While liquid surge does occur sometimes, it should be avoided as much as possible. You can have better control of the front, back or sideways surge of the liquid when you use baffles. Baffles have holes that allow the passage of liquid and prevent overflow.

(10) (B) You should expect a very high movement of the tanker when transporting liquid and lots of surging of the liquid.

If you use unbaffled tankers to transport liquid contents, expect the tanker to move a lot. Baffled tankers are better for transporting food products like milk.

(11) (B) Outage.

In order to provide an outage or room for liquid expansion, you should avoid loading a tanker to its brim.

(12) (C) Warm.

Liquids expand when warm, so there is a need for an outage. All liquids do not expand at the same rate or volume. Therefore, you need to know the outage rate before you load the liquid contents you are transporting.

(13) (D) Some acids are not dense liquids.

This is not true. Acids are very dense liquids. If the liquid is very dense, a completely filled tank may be beyond the lawful weight limitation. Therefore, if the liquid is dense, do not totally fill up the tank. You will know the liquid is dense by the quantity of expansion during transportation, the weight and if it meets the legal weight limitation.

(14) (D) All of the above.

You have to be very careful when driving tankers. Drive smoothly at all times because of the liquid surge and high center of gravity. Slow down, change lanes and stop smoothly. Do not take curves too fast. Take extra care to avoid turbulence.

(15) (D) All of the above.

To control liquid surge, you should apply consistent pressure on the brakes. Take it slow when you are stopping. Keep a good following distance between you and the vehicle in front. Increase the stopping distance if the road is wet. Always apply the brakes ahead of the stopping area. Apply stab or controlled braking to stop quickly and prevent a crash.

(16) (D) All of the above.

To avoid crashing, take it slow when you are stopping. Keep a good following distance between you and the vehicle in front. Apply stab or controlled braking to stop quickly and prevent a crash. Do not turn the wheel too fast while braking because the vehicle may slide. It may be more difficult to stop an empty tank than a filled one.

(17) (B) Skidding.

Too much acceleration or braking may make the vehicle skid. Skidding may make your tanker jackknife. When you drive slowly, you will be able to effectively avoid skidding.

(18) (D) None of the above.

Do not load, unload or drive until you have inspected the condition of the vehicle. You are performing this inspection to ensure the safety of the liquid or gaseous contents and to ensure safe driving.

(19) (D) All of the above.
Walk around and check for leaks beneath or in any part of the vehicle. It is very risky to load and drive a vehicle with leaks. It is also illegal. If you are caught, authorities will stop you from continuing with the drive, and you may be mandated to clean up any spill along the route you have been driving.

(20) (B) Before unloading.
When inspecting the valves, make sure there is an intake, discharge and shut off. Fix the valves in the most appropriate places before you start loading, unloading or driving the tanker. Also, inspect all the connections and pipes. Inspect the vents and covers of the manhole. Ensure there are available gaskets that keep the covers well closed.

(21) (D) None of the above.
All the answers are a kind of special purpose equipment. This includes grounding wires, bonding wires, the vapor recovery package and fire extinguishers. To ensure they are highly functional, always perform a practical inspection of the necessary equipment. Test if they are working before you take them along with you.

(22) (C) Supports the brakes.
When you apply brakes on either the long or steep downgrade of a mountain, it only supports the impact of the engine on braking. On the downgrade, you are expected to slowly reduce the speed of your vehicle.

(23) (C) Flash the emergency lights.
You should use emergency flashers to warn other drivers when there is trouble ahead, when you have to take a tight turn, when you are slowing down on the road and when you are stopping.

(24) (A) Hydroplaning.
Water or slush on the road sometimes leads to hydroplaning. The tires lose traction because they do not have enough contact with the road. This may affect your ability to steer your vehicle or apply the brakes. To regain control, let go of the accelerator and press the clutch. Do not use the brakes to reduce your speed.

(25) (C) Drive slowly in a gear you can control.
When going down a hill, reduce the speed to what you can easily control. Go easy on the brakes. Accelerating at a high speed down a hill will endanger the lives of everyone on board and the cargo. Lower the gear, or downshift.

7 Answers for Passenger Transport Test

(1) (C) They should not allow riders to carry common hazmats onto the bus.

Passenger drivers should not allow riders to carry common hazmats, such as car batteries or gasoline, onto a vehicle. Do a proper inspection because some of these hazmats might implicate you and everyone else if the police discover these materials in your vehicle.

(2) (C) Before loading and embarking on the journey.

Safety is a priority when driving a passenger vehicle. Before loading and embarking on the journey, go through the inspection report that the previous driver made to see what needs to be fixed or replaced. Inspect the parts of the vehicle and confirm that they are all highly functional. Ensure there has been proper repair or replacement of faulty or missing parts.

(3) (B) Refuse to sign it.

Go through the inspection report that the previous driver made to see what needs to be fixed or replaced. Refuse to sign the report if what needs to be repaired or replaced has not been taken care of.

(4) (A) Should always be inspected.

Before you drive, you should always inspect the reflectors and the lights. These are essential parts. Also, check the parking brake; the service brakes, which may include air hose couplings for vehicles with a semitrailer or trailer; the steering system; the horn; the tires; the wipers; the mirrors; coupling tools; the wheels and tires; and the emergency kits.

(5) (B) Closed.

Before you start driving, inspect the emergency exits and access panels and shut them if they are open. This is crucial to vehicle safety.

(6) (C) Fastened.

Before you load and drive your vehicle, check the interior and ascertain that it is safe for everyone to enter. Check to confirm the seats are safe and fastened for passengers and the driver. No seat should be loosened from the vehicle. This is very necessary because bumpy roads may displace and injure passengers if the seats are not fastened.

(7) (B) Always be turned on.

Keep the emergency light on at night. Also, make sure the inscription of "emergency exit" is very bold and clearly written on the emergency door.

(8) (D) All of the above.

Ensure the fire extinguisher is available and functional. Ensure there are spare parts, such as tires and electrical fuses. These components are vital in emergencies, such as fire outbreaks and mechanical and electrical faults.

(9) (A) Essential components.

Seat belts are essential components of the vehicle. They are meant to prevent passengers from jolting out of their seats. Inspect the seat belts in the vehicle. Always make use of yours and advise your passengers to use theirs. If there is wear and tear, replace the seat belts with new ones.

(10) (B) Informing riders of the existence of emergency exits.

You should always inform riders of the existence and location of emergency exits. This is standard procedure for loading a passenger vehicle.

(11) (C) Exit through the window or the closest door.

Riders can exit through the window or the closest door in an emergency. Make them aware of what to do when there is a fire outbreak or other unforeseen circumstances. Most times, it is not safe to stay on the bus when there is a dangerous situation.

(12) (D) Explosives should not be placed near passengers.

This statement is true. Explosives should not be placed near passengers. Transporting hazardous materials requires extra care. You are not allowed to transport hazardous materials if you do not have a hazmat license or endorsement. Ensure you are aware of what every cargo or baggage contains.

(13) (A) ORM-D.

Small arms and ammunitions are tagged as ORM-D. Buses are allowed to transport small-arm ammunition that is tagged ORM-D, emergency supplies for hospitals and drugs.

(14) (C) More than 100 pounds.

Solid poisons weighing more than 100 pounds should not be transported.

(15) (B) Not allowed.

Hazardous materials are threats to the health, property and safety of the people who are around. Therefore, you are not allowed to transport certain hazardous materials in passenger vehicles. This includes materials that are irritating to the nose and may result in health complications.

(16) (C) Standee line.

The standee line shows passengers where they can and cannot stand. Vehicles that are made to allow standing should have a marked line of two inches. The standee line helps keep passengers orderly and prevents falls.

(17) (D) All of the above.

When you are almost at a bus stop or final destination, announce the destination, state why you are about to stop, announce the time of departure before and/or after every stop and announce the number of the vehicle.

(18) (A) Ensure no one steps into the bus until it is time to move.

Ensuring no one steps into the bus until it is time to move will avoid theft and vandalism.

(19) (D) All of the above.

Before you begin a trip, clearly state the rules of no smoking, drinking, sexual intercourse, loud music and other rules. Be sure that passengers are aware of and understand each of the rules and, if possible, answer questions before you begin the trip. You are doing this so that everyone will be on the same page before you start moving.

(20) (D) All of the above.

Your responsibility as a driver continues while on the road. Always use the mirrors to supervise what the passengers are doing while you are driving. You may have to caution offenders and remind them of the rules. Correct those whose heads or arms are outside the windows. You must be vigilant.

(21) (B) Drop them off at the next bus stop.

If you encounter a belligerent rider, carefully handle the situation to ensure everyone is safe. Do not be in a hurry to drop off the belligerent rider until it is safe to do so. You may have to wait until the next stop before you drop the person off.

(22) (B) Wrong.

It is wrong to expect that other drivers will give you space after you have started pulling out. Do not assume. Wait for a gap before you merge to fill in the space.

(23) (B) Speed can lead to loss of life.

Slippery roads and speed are the greatest enemies to safety. They often result in the loss of lives and properties. Most roads are designed with an acceptable speed rate on the road. However, the speed limit of a car may not be appropriate for a bus. So you must be careful with the speed of your vehicle. Slow driving is far better than fast when taking curves.

(24) (C) 15 to 50.

When stopping at railroad crossings, apply the brake when your bus is about 15 to 50 feet away. Watch out for trains by listening and looking. If possible, step out of your seat to be sure if there are oncoming trains or not.

(25) (D) All of the above.

If the drawbridge does not have a traffic control attendant or green signal light, apply your brakes. You do not need to stop, but you have to slow down at least 50 feet from the drawbridge. Inspect to ensure the drawbridge is fully closed up.

8 Answers for School Bus Endorsement Test

(1) (C) Keep your gaze on the road.

Expect schoolchildren to create disturbances on the bus, but do not get distracted. While loading and unloading, your priority is to focus on what is happening outside. Do not ever take your gaze from the road to settle a dispute on the bus. You can use the mirrors to know what is happening at the back. If the dispute is intense, stop the vehicle and settle the conflict.

(2) (D) All of the above.

When managing students, observe the school rules and regulations on discipline before, during and after transit. Stop the bus as soon as you find a safe parking area. Turn off the vehicle and take your key as you step out of the vehicle. Remind the offender(s) of the rules and regulations.

(3) (A) Switch parties from their seats.

To ensure total resolution of the conflict, you may switch all parties from their seats. Place them where they cannot see or reach each other. You may even decide to instruct the offenders to sit closer to you for easier monitoring.

(4) (B) When getting off a bus or when boarding it.

Accidents are more likely to occur when students are boarding or getting off a bus. Following the proper procedures for loading and unloading will help avoid injuries or fatal incidents.

(5) (D) All of the above.

When approaching a designated bus stop, you must be very cautious with your speed. This area demands vigilance and skill. If you do not use caution, you may crash into pedestrians or other vehicles. You will have to use the mirrors, the lights, the signal arm and the crossing control arm. Watch both sides of your vehicle.

(6) (A) 5 to 10 seconds.

Before you arrive at the school bus stop, switch on the warning lights a few seconds (5 to 10) or 100 to 500 feet ahead.

(7) (A) 3 to 5 seconds.

Do not just pull over without switching on your indicator. Switch on the right-side indicator three to five seconds before you are about to pull over. This will help prevent accidents by telling drivers of your next action so they can react accordingly.

(8) (C) Walk.

Stop the school bus 10 feet from the bus stop. This will make the students walk toward the bus. You will also be able to count the students.

(9) (A) Neutral or park.

Set the transmission in neutral or park. It should be in neutral if a park shift point does not exist. This will prevent the vehicle from moving. Also, remember that you have to wait for traffic to totally clear before you open the door and ask the students to start coming in.

(10) (C) Wait for the school bus to arrive at a specific bus stop.

Students are instructed to wait for the school bus to arrive at a specific bus stop. This location should be where the driver can see the students when approaching the bus stop. The students should not wait anywhere else. They should arrive on time because the driver will not wait longer than necessary.

(11) (D) All of the above.

No students should enter the bus until the driver has signaled them to come in. Remember to take a count of the students. You can learn the names of the students, and if the number of students on board does not tally with the count or the expected number, ask for the whereabouts of the missing student(s).

(12) (C) In a single-file line.

Ask the students not to rush in. Instead, they should be in a single-file line and use the handrails. Turn on the dome light if the loading area is dark. Avoid being hasty when loading. Ensure everyone is seated and ready to be transported before embarking on the trip.

(13) (C) When everyone is seated.

Do not drive until everyone is seated. If you do so, students who are still standing may stumble and get injured.

(14) (D) None of the above.

If the student(s) will cross the roadway, step down from the bus and walk about 10 feet from the bus to where you can see the students crossing. Stay at the right side of the road, where you can clearly view the students' feet. Look in all directions and ensure the road is clear for crossing.

(15) (B) Row by row.

When the students leave the bus row by row, there will be less chance of an accident. Teach them how to come down from the bus in an organized way. Also, when it seems like you have fully unloaded, check inside the bus to be sure that there are not any students who have been left behind.

(16) (B) Switch them off.

After you have accounted for each student, prepare to leave by switching off the alternating flashing lights, shutting the door, putting on your seat belt, turning on the engine, running the transmission, letting go of the parking brake, switching on the left turn signal and looking at the mirrors again. Then drive away.

(17) (C) A danger zone.

The left zone of a bus is called a danger zone because of vehicles that are passing by. The danger zone of a school bus is an area where schoolchildren are most likely to be injured by a vehicle.

(18) (D) All of the above.

To avoid accidents when driving, you need to properly adjust the mirrors, which are very useful for looking out for traffic, objects and the welfare of students.

(19) (A) The outside flat mirrors.

The outside flat mirrors at the left and right sides of your vehicle are used to check incoming vehicles and activities going on outside the back of the vehicle. A blind spot is directly in front and under each mirror, as well as the back of the rear bumper. The distance of the blind spot is about 50 to 150 feet; sometimes it goes as far as 400 feet.

(20) (C) The sides of your vehicle.

When you properly adjust the outside flat mirrors, they will help you see the sides of the bus.

(21) (A) The outside flat mirrors.

When you properly adjust the outside flat mirrors, they will help you have a view of approximately 200 feet or the length of four buses behind your bus, the sides of the bus and the back tires where they contact the ground.

(22) (C) They are exact in reflecting the size of people.

This is not true. The outside convex mirrors provide a panoramic view of what is happening at the left and right sides outside. Like the flat mirrors, they give you a view of the students who are at the side, traffic and other ongoing activities. These mirrors will not give you the exact distance and size of objects and people.

(23) (D) The outside crossover mirrors.

The outside crossover mirrors are on the left and right front sides of your vehicle. These mirrors give you a good view of danger zones in front of and outside the bus. These zones are hard to see with the eyes only. With the mirrors, you will have a proper view of the front wheel and the service door areas.

(24) (D) All of the above.

Occasionally view the flat, convex and crossover mirrors sequentially to keep everyone safe from the danger zones. You should not substitute one mirror for the others. Also, remember to complement the mirrors with your actual eyes.

(25) (A) The overhead rearview mirror.

The overhead rearview mirror is suspended above the windshield in the driver's area. It allows you to check what is going on inside the bus without having to turn your head around and take your concentration totally off the road.

9 General Knowledge Test

1. Braking very hard or locking the wheels is also known as ________.

A. Brake down

B. Overbraking

C. Break up

D. Oversteering

2. How does alcohol affect driving?

A. It results in bad judgment.

B. It makes the driver speed up or drive too slow.

C. The driver will drive in the wrong lane.

D. All of the above.

3. How do you test your parking brake's functionality?

A. Increase the speed to 10 mph and set the parking brake.

B. Double the speed after applying the parking brake.

C. Turn on the parking brake and slowly pull against it.

D. Decrease the speed and slowly park at a roadside.

4. What do you do to a burning tire?

A. Remove the tire.

B. Cover it with black cloth.

C. Pour water on it.

D. All of the above.

5. Professional drivers look at least how many seconds ahead?

A. 10 to 20

B. 12 to 15

C. 20 to 35

D. 40 to 50

6. The size of your vehicle ____________.

A. Does not affect drivers behind you

B. Makes it difficult for drivers at the rear to have a clear sight of hazards in front

C. Makes drivers in front slow down

D. Increases the speed of drivers behind you

7. What should you do when the spring brakes are on?
A. Turn on the accelerator
B. Increase the brakes
C. Not press the brake pedal
D. All of the above

8. As a cargo driver, it is your duty to do which of the following?
A. Inspect your cargo.
B. Identify weight that is not balanced.
C. Know if your cargo is well secured.
D. All of the above.

9. When you are not driving fast, 12 to 15 seconds equals __________.
A. 1 block
B. 2 blocks
C. 12 blocks
D. 15 blocks

10. Which of these affects brake fading?
A. Adjustments
B. The steering
C. The rear tires
D. None of the above

11. When making lane changes, you should _________.
A. Make sure the passengers are awake.
B. Check your mirrors.
C. Drive when there is someone in your blind spot.
D. None of the above.

12. What should drivers do when driving trailers?
A. Change lanes quickly.
B. Push the brakes abruptly when stopping.
C. Avoid moving the steering wheel suddenly.
D. Reduce the following gap between them and other vehicles.

13. Before starting a trip, which of the following methods can prevent drowsiness?
A. Proper sleep
B. Not taking sleep-inducing medicines
C. Avoiding late-night driving
D. All of the above

14. When should you review the last driver's trip report?
A. After every trip
B. Once every week
C. During a trip
D. Twice a month

15. Which of the following statements is true about fatigue?
A. It should be addressed only when it becomes critical.
B. It is highly welcome because it improves our fight-or-flight mode when driving.
C. It is a factor that may lead to an accident.
D. All of the above.

16. A skid occurs under which of the following circumstances?
A. When the tires are new
B. When the tires are old
C. When the tires do not have a grip on the road
D. When the brake is depressed in an emergency

17. What should you do when backing up a trailer?
A. Stay off the brakes.
B. Position your vehicle for curves.
C. Turn the steering wheel opposite the direction of the turn.
D. Apply the emergency and parking brakes together.

18. When approaching a turn, unsafe coasting occurs under which of the following conditions?
A. When the ABS is turned off
B. If you change gears smoothly
C. When the gear shift is in neutral
D. When your vehicle is not in gear

19. What does rust around the wheel mean?
A. The nuts are very tight.
B. The wheels have just been welded.
C. The nuts are not well tightened.
D. The weather is bad.

20. If you must drive in a deep puddle, you should do which of the following?
A. Slow down and set the transmission in a low gear.
B. Start slowly and speed up in the next 15 seconds.
C. Drive out of the puddle quickly.
D. None of the above.

21. Which of these is true about wet roads?
A. They often reduce stopping distance.
B. They do not affect stopping distance.
C. They make the stopping distance twice as long as normal.
D. None of the above.

22. Which of the following is supplying too much force to the wheels and making them spin?
A. Overbraking
B. Oversteering
C. Overaccelerating
D. Overreacting

23. Which of the following is true of driving very fast?
A. It often leads to delays.
B. It may lead to skidding.
C. It is sometimes necessary.
D. All of the above.

24. Which of these is not a basic method for shifting up using manual transmission?
A. Let go of the accelerator, press the clutch and move to neutral.
B. Let go of the clutch.
C. Never release the clutch.
D. Allow the engine to slow down to what the RPM needs.

25. When you see people working on any part of the road, you should consider it ________.
A. A hazard
B. A time to stop immediately and turn back
C. Road rage
D. An opportunity to stop and walk around

26. Which of the following is not a hazard?
A. Passengers in the vehicle
B. Foreign objects
C. Ice cream trucks
D. Pedestrians

27. You are driving on the road and suddenly hear a loud bang. What should you do?
A. Hold on to the steering wheel tightly.
B. Do not take your foot off the brake.
C. Inspect the front and rear tires.
D. All of the above.

28. What is BAC?
A. Back Acceleration Contact
B. Blood Alcohol Concentration
C. Built Adapter Control
D. Bus Acceleration Control

29. If you are driving and the steering starts feeling heavy, what may be the reason?
A. Brake failure
B. Tire failure
C. Air tank failure
D. Power failure

30. The thumping or vibration of a vehicle is a sign of which of the following?
A. The rear tire is flat.
B. You need more fuel.
C. The service brake is malfunctioning.
D. There is danger ahead.

31. Which of these is not a determinant of BAC?
A. Quantity of alcohol consumed
B. How fast alcohol is consumed
C. The person's weight
D. The person's intellect

32. Fishtailing is a result of ________.
A. Driving in swamp water
B. Failure of a back tire
C. Too much water in the storage tank
D. None of the above

33. The Department of Transportation has mandated which of the following?

A. All tractors that were manufactured on or after March 1, 1997, should have an antilock braking system.

B. All converter dollies, trucks, buses and trailers that were manufactured on or after March 1, 1998, should have antilock braking systems.

C. Hydraulic brake trucks that weigh 1,000 pounds and were produced on or after March 1, 1999, should have an ABS component.

D. All of the above.

34. How will you know if your vehicle has ABS?

A. A bus, tractor or truck will have yellow ABS lights at the device component.

B. There will be a red ABS light on the right side at the front or the back if it is a trailer.

C. The light turns off when you are underway.

D. All of the above.

35. Which of the following does high BAC affect?

A. Muscles

B. Control

C. Coordination

D. All of the above

36. Which of the following should be done before you drive?

A. Get sufficient rest and sleep.

B. Eat a healthy meal.

C. Visit your doctor.

D. All of the above.

37. What should you do if you notice that there is a fire on the vehicle?

A. Send for help.

B. Use a fire extinguisher.

C. Keep the trailer doors shut.

D. All of the above.

38. Which of the following is a cause of fire?

A. Driving on slippery surfaces

B. Fueling at fuel stations

C. Rays of the sun on the windscreen

D. Smoking and driving

39. What should you do if you have a load that is reducing your speed?
A. Keep on driving in your lane.
B. Stay in the right lane.
C. Stay in the left lane.
D. Drive in third gear.

40. What does a bridge formula do?
A. It allows the lower maximum weight in axles that are not far apart.
B. It gives the exact distance of a bridge.
C. It reduces the stress of driving over bridges.
D. All of the above.

41. Which of the following will occur if the vehicle you used for the skills test does not have air brakes?
A. You will be restricted to vehicles with air brakes.
B. You will have to take the test again.
C. You will have to take a combination vehicle test.
D. You will be allowed to drive vehicles that do not have air brakes.

42. What should you do when you are feeling sleepy when driving?
A. Stop and take a rest.
B. Do your best to stay awake.
C. Play some music.
D. Concentrate harder on the road.

43. Why should you be aware of your vehicle's size and weight when you are crossing or entering traffic?
A. Because big vehicles accelerate slowly and need lots of room
B. Because the speed rate of acceleration varies in every load
C. Because you need to ensure you can get across traffic before you enter the road
D. All of the above

44. Which of the following is the application of brakes without stopping and letting go when the wheels are shut up?
A. Stab braking
B. Controlled braking
C. Emergency parking
D. Service parking

45. How do you cool the engine of your vehicle in hot weather?
A. Ensure there is sufficient engine oil.
B. Ensure there is sufficient antifreeze and water.
C. Check if the V belt is tight.
D. All of the above.

46. Which of the following is an example of an explosive?
A. Helium
B. Dynamite
C. Oxygen
D. Pesticide

47. Where does hydroplaning often occur?
A. Hard surfaces
B. Hot surfaces
C. Slippery surfaces
D. Muddy surfaces

48. What should you do when you are at an intersection?
A. Quickly move through it.
B. Accelerate the vehicle.
C. Adjust the gears while proceeding.
D. None of the above.

49. Which of the following may lead to hydroplaning?
A. Slush
B. Tar
C. Engine oil
D. Sun

50. You are driving in winter and notice that the temperature of the weather is below freezing. You should look out for which of the following?
A. Black ice
B. Melting ice
C. Snow
D. All of the above

10 Air Brakes Test

1. An air brake needs which of the following to work?
A. Compressed pressure
B. Light pressure
C. Light gear
D. Free air

2. What does the generated pressure from the compressed air do to the brakes?
A. It smooths the brakes.
B. It makes it difficult to drive.
C. It limits the friction of the brakes.
D. It creates friction against the brakes and the tires.

3. When should you use the service brake?
A. When pulling over
B. When driving
C. When the brake system is malfunctioning
D. All of the above

4. When is the hydraulic braking system used?
A. When there is a need to park
B. When the vehicle needs water
C. When the brake system is functioning effectively
D. When there is a failure of air supply to the brake chamber

5. When does pneumatic energy become mechanical energy?
A. When the brake ceases to function
B. When the neural system of the body reacts to the mechanical system of the vehicle
C. When electrical energy converts to mechanical energy
D. When compressed air makes the brake system functional

6. At an optimal level of about 125 PSI, which of the following makes sure there is continual pumping of air from the compressor?
A. The storage tank
B. The gauge
C. The gear
D. The air compressor governor

7. When air pressure mounts up to the air compressor governor cut-off, what is the normal air loss rate for a combination vehicle if the engine is turned off and you have fully pushed the foot brake?

A. Below 2 PSI in 1 minute

B. Above 3 PSI in 1 minute

C. Below 5 PSI in 2 minutes

D. Below 3 PSI in 1 minute

8. Which of the following is false about storage tanks?

A. They cause the brakes to be applied as much as possible.

B. They store air from the compressor.

C. They are a very vital component of the air brake system.

D. None of the above.

9. You are on a highway and a puppy suddenly runs into the road. Which of these components should you use to save the day?

A. The mirrors

B. The brake pedal

C. The tires

D. The air compressor governor

10. Which of the following statements is true about the safety valve?

A. It keeps the air in the storage tank.

B. It extinguishes fires.

C. It ejects air when the pressure gets to its maximum point.

D. All of the above.

11. Which of the following is another name for a brake line?

A. Stop line

B. Supply line

C. Stop brake

D. Restricted brake

12. Which of the following is false about the front brake limiting valve?

A. It has both normal and slippery modes.

B. It limits the flow of pressure.

C. It allows your vehicle to stop normally.

D. It is a limitation to the back brake.

13. What is the function of the air tank draining valve in the air brake system?
A. It removes water and oil residuals from the tank.
B. It drains the solid components from the air brake system.
C. It keeps substances from draining out of the air brake system.
D. It freezes engine oil in the air tank.

14. Air brake systems have a/an __________.
A. Alcohol cooler
B. Alcohol evaporator
C. Pressure evaporator
D. Pressure cooler

15. The slack adjuster will twist __________ when you press the brake pedal.
A. The wedge brake
B. The shaft of the brake cam
C. The steering screw
D. None of the above

16. A disc brake applies air pressure against the __________.
A. S-cam
B. Brake chamber
C. Wedge brake
D. Compressor

17. Which of these components shows you the exact amount of air pressure you have exerted on the brakes?
A. The supply pressure gauge
B. The application pressure gauge
C. The display pressure gauge
D. The electronic pressure gauge

18. Which of the following tells you when your pressure supply is almost below 60 PSI?
A. Low air pressure signal
B. Red light
C. Buzzer
D. All of the above

19. When you have ABS, which of the following will apply?
A. Your wheels will not lock.
B. Your driving pattern will change.
C. There will be no need for normal brakes.
D. All of the above.

20. What happens when your ABS is malfunctioning?
A. You should continue driving normally.
B. Your normal brakes are still available.
C. Nothing will change in the speed of your vehicle.
D. All of the above.

21. How do you perform a normal stop with an air brake system?
A. Start by pulling down the brake pedals.
B. First, switch off the ignition.
C. Begin with the brake pedals.
D. All of the above.

22. You can change how you brake when driving which of the following?
A. A straight truck or combination vehicle that has ABS on both axles
B. A truck with ABS
C. A tractor with ABS
D. A bus with ABS

23. Which of the following techniques allows you to apply the brakes very hard without closing up?
A. Parking
B. Controlled
C. Emergency
D. Service

24. Which of the following statements is false about air brake systems?
A. They are only used by road vehicles.
B. It is very difficult to store high atmospheric pressure in a storage tank.
C. Braking is not effortless when using air brakes.
D. All of the above.

25. What do you do if the compressor uses a belt to function?
A. Do not concern yourself with its maintenance.
B. Check for tightness.
C. Allow the belt to loosen.
D. Detach and fix it.

11 Doubles and Triples Test

1. Which of the following statements is not true about doubles and triples?

A. You use the same speed you use with combination vehicles to prevent a rollover.

B. You need to drive slowly on ramps.

C. The knowledge of driving singles should not be directly transferred to the driving of combination vehicles.

D. None of the above.

2. Doubles and triples are _________.

A. Small

B. Long

C. As big as four combination vehicles

D. Not suitable for highways

3. Which of the following is true about driving doubles and triples?

A. You should not be in a hurry to cross traffic.

B. You should first confirm if the space is long enough for your double or triple vehicle.

C. You should maintain a good distance.

D. All of the above.

4. Which of the following often leads to road accidents?

A. Dry roads

B. Asphalt roads

C. Gravel roads

D. Slippery roads

5. What is the consequence of careless parking?

A. Breakdown of the vehicle

B. Difficulty in pulling out

C. Too much space

D. Overheating of the engine

6. If you want air to flow to all parts of the trailer, what should you do?

A. Engage the parking brake and press the red switch.

B. Let go of the trailer's hand brake.

C. Close the emergency line valve.

D. Smell for fuel leaks.

7. Which of the following is true about the flow of air?
A. It should flow to the rear only.
B. There is no sound when the air is released from the air line.
C. Air must go to all areas.
D. It is only needed in the engine compartment.

8. Which of the following is not a step in inspecting the emergency brakes?
A. Keep the trailer emergency brake switched off.
B. Power up the air brake system.
C. Ensure there is no obstruction to the rolling of the trailer.
D. Slowly pull out the trailer.

9. Which of the following is not a step in inspecting the service brakes?
A. Keep the air pressure normal.
B. Gently drive the vehicle to the front.
C. Press the trailer's brakes using the trolley valve or hand control.
D. Use the wheel to inspect the trailer's brakes.

10. What is a converter dolly used for?
A. Coupling
B. Brake application
C. Energy conversation
D. All of the above

11. If the rear trailer does not have a spring brake, what should you do?
A. Bring the tractor and the trailer closer.
B. Integrate the emergency line.
C. Charge up the air storage tank.
D. All of the above.

12. What should you do when using the front semitrailer and the tractor to pick up the converter dolly?
A. Set the combination very close to the converter dolly.
B. Lock the pintle hook.
C. Protect the dolly in a lifted position.
D. All of the above.

13. Which of the following is not involved in joining the converter with the first trailer?
A. Put the first semitrailer in front of the dolly tongue.
B. Connect the front trailer with the dolly.
C. Connect the emergency line and the service line together.
D. Safely set the converter gear in a lifted position.

14. What should be the height of the trailer when coupling?
A. A bit below the middle of the fifth wheel
B. Higher than the fifth wheel
C. Exactly 10 inches high
D. None of the above

15. Which of the following is true about the air brake system of combination vehicles?
A. The brakes will not work if there is no air supply and storage.
B. It needs water more than air.
C. It should be serviced once every decade.
D. All of the above.

16. You are in the process of uncoupling the back trailer, and your vehicle does not have a spring brake. You should ______________.
A. Open the first trailer wheels.
B. Lock up the second trailer wheels.
C. Open the air shut-offs.
D. None of the above.

17. To remove the weight or pressure from the dolly, you should release the rear semitrailer's _____________.
A. Shut-off valve
B. Landing gear
C. Air compressor
D. Electrical lines

18. Which of the following is true about converter dollies?
A. They come with spring brakes.
B. They convert kinetic energy to nuclear energy.
C. You should keep the landing gear up when uncoupling.
D. All of the above.

19. You should ______ the pintle hook when uncoupling converter dollies.

A. Latch

B. Unlatch

C. Hook

D. None of the above

20. The flinging of the dolly bar makes it difficult to _______ again.

A. Uncouple

B. Couple

C. Unbolt

D. Drive

21. How do you uncouple a triple trailer rig?

A. Push in the dolly.

B. Pull out the dolly.

C. Raise the landing gear.

D. All of the above.

22. When inspecting doubles and triples, you should make sure there is no fault in ________.

A. The glide plate

B. The air lines

C. The kingpin

D. All of the above

23. You should apply which of the following to the system area of your doubles and triples?

A. Grease

B. Water

C. Acid

D. Peroxide

24. Which of the following is not a step in inspecting and maintaining doubles and triples?

A. Only fix the emergency air line.

B. Keep the spare tire safe.

C. Ensure proper locking of the pintle.

D. Provide security for the safety chains.

25. The pintle of doubles and triples should be ______ during an inspection.

A. Locked

B. Unlatched

C. Left

D. Up

12 Hazardous Materials Test

1. Which of the following statements is not true about hazmats?
A. They are also called hazardous materials.
B. They are toxic.
C. They are risky only to the health of the driver.
D. None of the above.

2. A commercial driver license is __________.
A. Compulsory for every driver transporting hazardous materials
B. Mandated only for drivers transporting hazardous materials out of state
C. Required only for drivers of combination vehicles
D. All of the above

3. Helium, oxygen and propane are examples of which class of hazardous materials?
A. Class 1
B. Class 2
C. Class 3
D. Class 4

4. Hydrogen peroxide and ammonium nitrate are examples of which class of hazardous materials?
A. Class 2
B. Class 3
C. Class 5
D. Class 7

5. Which of the following are examples of radioactive substances?
A. Hydrogen peroxide and ammonium
B. Plutonium and uranium
C. Battery acid and hydrochloric acid
D. Asbestos and charcoal

6. Hazmat regulations are also known as ________.
A. Hazardous storage
B. Containment rules
C. Harmless instructions
D. Regulations and rules

7. Which of the following is a result of breaking hazmat rules?

A. Imprisonment

B. Fatal accident

C. Fines

D. All of the above

8. If the capacity of the portable tank is above 1,000 gallons, the size of the shipping name should not be less than how many inches in height?

A. 2

B. 4

C. 6

D. 3

9. Which of the following is a type of bulk package that does not need a shipping name?

A. Large cargo tank

B. Intermediate bulk container

C. Large bulk container

D. None of the above

10. The vehicle should be ______ when loading and unloading.

A. Static

B. Moving

C. Empty

D. Marked

11. Oxidizers should be _________.

A. Packed tightly

B. Allowed to slide

C. Put in an oxygen tank

D. Given less attention

12. Smoke around hazardous materials is __________.

A. Required to be five feet away

B. Prohibited

C. Not a danger

D. None of the above

13. A portable tank __________.

A. Is fixed on the vehicle

B. Is unloaded and loaded before it is placed on the vehicle

C. Does not show the owner's name

D. Is the same as a cargo tank

14. Which of the following cargo tanks is used for hazardous gases?

A. MC306

B. MC321

C. MC331

D. MK40

15. The shipper is in charge of which of the following?

A. Identification, classification, grouping and packaging of products

B. Obtaining shipping papers

C. Transporting goods to different points

D. All of the above

16. The shipping papers are ___________.

A. A tool for communication

B. A green paper

C. An optional paper

D. All of the above

17. Where should drivers keep hazardous material shipping papers?

A. In a bag on the driver's door

B. On the driver's seat

C. Within arm's reach of the driver

D. All of the above

18. Which of the following is used to signal other drivers of the presence of hazardous materials?

A. Placards

B. Green lights

C. Yellow lights

D. Traffic lights

19. Where should placards be placed on a vehicle carrying hazardous materials?

A. Inside

B. Outside

C. Under

D. On the windshield

20. A vehicle that has been placarded should have no fewer than how many placards?
A. Two
B. Three
C. Four
D. Five

21. Placards are placed __________.
A. Only at the front
B. On the sides
C. On the front, back and sides of the truck
D. On top of and below the truck

22. Which of the following is true about corrosive liquids?
A. They should not be close to explosives, blasting agents, oxidizers or poisonous gases.
B. They can be close to heat sources.
C. You should load the flammable liquid while the engine is on.
D. All of the above.

23. What should you do if the hazardous materials you are carrying are susceptible to combustion?
A. Provide proper ventilation.
B. Ensure all holes are closed.
C. Increase the speed of your vehicle.
D. All of the above.

24. Which of the following is true of cylinders containing hazardous materials?
A. They should be stored in racks.
B. They can be kept in a horizontal position.
C. They may be kept right-side up.
D. All of the above.

25. A cargo tank is ___________.
A. A type of bulk packaging
B. A nonpermanent appendage
C. A permanent appendage
D. All of the above

13 Combination Vehicles Test

1. Piling too much cargo into a combination vehicle will affect __________.

A. Its stability

B. Its size

C. Its durability

D. All of the above

2. You should leave at least __ second (s) for every __ feet of the length of your vehicle at a speed under 40 mph.

A. Two seconds; ten feet

B. Three seconds; five feet

C. One second; ten feet

D. Two seconds; fifteen feet

3. Early braking is possible under which of the following conditions?

A. The vehicle is fairly small.

B. The weight of the vehicle is reduced.

C. There is an increase in the following distance.

D. The vehicle is driving in a dry season.

4. What is cheating?

A. When you wrongly overtake other vehicles

B. When the front and rear wheels are in different positions when you are going around a corner

C. When you fail to stop when the traffic light is red

D. None of the above

5. The trolley valve is also called __________.

A. The trailer hand valve

B. The basket valve

C. The tractor valve

D. All of the above

6. Which of the following is used to test the brakes?

A. The Johnson bar

B. The trailer hand valve

C. The trolley valve

D. All of the above

7. Parking with the trailer hand valve might lead to __________.

A. A nonstop movement

B. Air leaks

C. Overbraking

D. Fuel shortage

8. What happens when the tractor protection valve is closed?

A. Air will start leaking.

B. It will prevent air from escaping.

C. Air will go through the service line.

D. All of the above.

9. How do you identify the air supply control valve in modern vehicles?

A. It is green.

B. It has a red eight-sided knob.

C. It has blue squares around it.

D. It buzzes when the vehicle is in transit.

10. Which of the following is used to transfer air around the air brake system?

A. The parking air line

B. The transfer air line

C. The service air line

D. The circulatory air line

11. The emergency air line is also known as the ___________.

A. Control line

B. Supply line

C. Signal line

D. Urgency air line

12. What color is the service line?

A. Green

B. Yellow

C. Blue

D. Red

13. Which of the following are dead ends, which are connected with the hose when they are idle?
A. Glad hands
B. Hose couplers
C. Rear compartment
D. Dummy couplers

14. Which of the following statements is true about the shut-off valves?
A. They prevent the closure of the air lines when there is no towing trailer.
B. They shut off the engine when parking.
C. They allow air line closure when there is no towing trailer.
D. They shut off the fuel tank in emergencies.

15. In coupling a tractor-semitrailer, what should you do when observing the area?
A. Check if the vehicle area is clear.
B. Ensure the spring brake is on and the wheel is chocked.
C. Check if the cargo is intact and not susceptible to movement.
D. All of the above.

16. When coupling combination vehicles, what will happen if the trailer is very high?
A. The load will topple over.
B. Coupling may become difficult.
C. The steering wheel will stiffen.
D. All of the above.

17. The emergency line should be connected to the _________ glad hand(s).
A. Emergency
B. Service
C. Parking
D. Emergency and service

18. If there is an interlock between the fifth wheel and the kingpin, you should ___________.
A. Stop immediately.
B. Look under the trailer.
C. Steer toward the road.
D. Apply the reverse gear.

19. Which of the following is not true about lifting the landing gear?
A. You should start from the lowest point.
B. You should increase the gear range when the landing gear becomes weightless.
C. You should lift the gear and stop midway.
D. All of the above.

20. Which of the following is not a step in uncoupling tractors/semitrailers?
A. Set the rig in the right position.
B. Reduce the pressure against the locking jaw.
C. Chock the wheels to prevent movement.
D. None of the above.

21. Why should you switch off the air supply when reducing the pressure against the locking jaw?
A. To increase the air supply
B. To prevent a fire outbreak
C. To generate compressed air
D. To lock the brakes

22. When uncoupling, the fifth wheel should be ________.
A. Lifted
B. Down
C. Used only when the weather is hot
D. Used only when the weather is cold

23. Lifting the fifth wheel will __________.
A. Reduce the weight of the vehicle
B. Make the vehicle noiseless
C. Make the recoupling easier
D. All of the above

24. After lifting the release handle lock, what should you do when uncoupling trailers?
A. Increase the landing gear.
B. Switch off the vehicle.
C. Inspect the rear tires.
D. Place the release handle lock in an open position.

25. Which of the following is not a step in testing the trailer emergency brakes?
A. Power up the air brake system.
B. Prevent the trailer from rolling.
C. Put the air supply control on emergency.
D. Pull out the tractor protection valve.

14 Tanker License Test

1. A tanker license is often required for cargo tanks with a capacity of _____ gallons.

A. 200 or more

B. 1,090 or more

C. 119 or more

D. 15 or more

2. Before you drive, you should do which of the following to your vehicle?

A. Load

B. Unload

C. Inspect

D. Uncouple

3. Tankers often have ________ when driving.

A. Air pressure

B. Liquid turbulence

C. Low center of gravity

D. All the above

4. When driving a tanker, you should remember which of the following?

A. The vehicle can easily go off the road.

B. The top of the vehicle is very light.

C. Liquid hardly rolls down.

D. Rollovers do not occur in curves.

5. When the tank is nearly filled to the brim, which of the following is likely to occur when the vehicle is moving?

A. Liquid surge

B. Fuel leaks

C. Air leaks

D. All of the above

6. What can happen when you park a tanker on a slippery surface?

A. The wave might move the tanker in its direction.

B. The tanker will drive itself.

C. You can drive the tanker the way you would drive a school bus.

D. You do not need to worry about liquid overflow.

7. The bulkhead is used for __________.

A. Carrying bulk loads

B. Partitioning liquid tanks

C. Piling loads in one direction

D. All of the above

8. Liquid tanks with bulkheads are also known as ___________.

A. Fluid tankers

B. Liquid tankers

C. Baffled tanks

D. Portable tanks

9. Baffled tanks help you have control of the _______ of the liquid.

A. Front

B. Back

C. Sides

D. All of the above

10. Unbaffled tankers are also known as __________.

A. Smoothbores

B. Unbalanced tankers

C. Balanced tankers

D. Baffled tankers

11. Outage is ___________.

A. Room for expansion

B. Space through which air escapes

C. The hole through which water escapes

D. The point where power is plugged in

12. Which of the following is very important when loading liquids?

A. The expansivity rate

B. The outage rate

C. Expansion

D. All of the above

13. How do you determine the quantity of liquid you can load?
A. By the quantity of the liquid expansion during transportation
B. By how much the liquid weighs
C. If it meets the legal weight limit
D. All of the above

14. When do you need to drive smoothly?
A. When slowing down
B. When changing lanes
C. When stopping
D. All of the above

15. Which of the following is not a step in controlling liquid surge?
A. Decreasing stopping distance
B. Slowing down when stopping
C. Not steering too fast
D. None of the above

16. Driving your tanker too fast will _________.
A. Prevent a crash
B. Make it tip over
C. Increase the stopping distance.
D. Prevent liquid surge

17. Overbraking may lead to which of the following?
A. Higher speed
B. Acceleration
C. Skidding
D. All of the above

18. Why should you inspect your tanker?
A. For safe driving
B. For the safety of the gaseous contents
C. For the safety of the liquid contents
D. All of the above

19. Which of the following is not true about driving a vehicle with leaks?
A. Drivers should know how.
B. It is risky.
C. Authorities may stop you.
D. All of the above.

20. The vents should be __________.

A. Free from particles

B. Filled with particles

C. Filled with leaks

D. None of the above

21. Which of the following is not true about special purpose equipment?

A. It is always labeled red.

B. It should be inspected before use.

C. It must be highly functional.

D. An example is bonding cables.

22. Which of the following is not a type of vehicle inspection procedure for a tanker?

A. Pre-trip inspection

B. During the trip inspection

C. After the trip inspection

D. Semi-inspection

23. The ABS system does which of the following?

A. Helps reduce the brake pressure

B. Prevents your wheels from locking themselves

C. Helps increase the pressure in the tires

D. Helps prevent leakage of air

24. Escape ramps are made to ___________.

A. Safely reduce the speed of runaway vehicles on a steep mountain

B. Make driving more enjoyable

C. Increase the level of acceleration

D. All of the above

25. Why should you inspect your tanker?

A. For safety

B. For maintenance

C. For prevention of road accidents

D. All of the above

15 Passenger Transport Test

1. Riders carrying common hazmats like gasoline should be __________.

A. Allowed into your vehicle

B. Banned from your vehicle

C. Charged higher fees

D. Charged lower fees

2. Which of the following is true about the inspection of passenger vehicles?

A. You should go through the inspection report that the previous driver made to see what is lacking.

B. You should inspect the parts of the vehicle.

C. You should ensure every object and person you are driving is safe before loading and embarking on a trip.

D. All of the above.

3. Which of the following is true about the inspection report?

A. You should never sign an inspection report.

B. It is wrong for a driver to go through an inspection report.

C. Your signature is mandatory on the inspection report.

D. None of the above.

4. Which of the following needs inspection before starting a trip?

A. The wipers

B. The mirrors

C. The emergency kits

D. All of the above

5. Which of the following do you inspect before you load your vehicle?

A. The interior

B. The railing

C. The floor covering

D. All of the above

6. The emergency window should be __________ when driving.

A. Closed

B. Open

C. Curtained

D. None of the above

7. How should the word "emergency exit" be written?
A. Stylishly
B. Boldly
C. On the regular door
D. On the top of the vehicle

8. You should ensure the fire extinguisher of your passenger vehicle is ________.
A. Functional
B. Available
C. Filled
D. All of the above

9. Seat belts are to be worn by _________.
A. Passengers only
B. Drivers only
C. Drivers and little children
D. Drivers and passengers

10. Keeping baggage by the doorway should ___________.
A. Be allowed
B. Be forbidden
C. Incur an extra fee
D. Be moderated

11. Smooth movement within the vehicle is ________.
A. Not possible in passenger buses
B. A possibility in passenger buses
C. Achievable only if there are very few passengers
D. A cause of delays

12. In a passenger bus, you are not allowed to transport ___________.
A. Poisonous gas
B. Liquid poison
C. Irritating materials
D. All of the above

13. Which of the following are you allowed to transport on a passenger bus?
A. Battery cells
B. Solid poisons weighing more than 100 pounds
C. ORM-D
D. Liquid poisons

14. Which of the following is true of explosives?
A. They should be kept in front of passengers.
B. They should not be kept near passengers' seats.
C. They should be kept under passengers' seats.
D. None of the above.

15. Which of the following statements is true about loading gasoline or battery cells in passenger vehicles?
A. They should not be allowed.
B. They are permissible.
C. They should be loaded fully.
D. They are not hazardous materials.

16. The standee line is often approximately how many inches?
A. 2
B. 3
C. 4
D. 5.

17. What should you do when arriving at the final destination?
A. Announce the name of the vehicle.
B. Repeat the name of the previous bus stops before announcing the last bus stop.
C. Announce the time you are stopping.
D. All of the above.

18. Which of the following should you do to avoid stealing and vandalism?
A. Do not stop unless there is a checkpoint nearby.
B. Keep everyone outside until you are ready to move.
C. Employ a security officer.
D. All of the above.

19. Your responsibility for ensuring safety as a driver __________.
A. Continues while on the road
B. Ends after the beginning of the trip
C. Is possible only at the beginning and end of your journey
D. Is for the first trip only

20. When you're at a bus stop, __________.
A. Some passengers may fall when entering or exiting the bus.
B. You should always remind passengers to be cautious of their steps to avoid stumbling.
C. You should wait for every rider to be seated comfortably before you depart.
D. All of the above.

21. Most bus crashes occur at/in ________.
A. Straight roads
B. Zebra crossings
C. Intersections
D. Traffic

22. Which of the following is true about joining traffic after a stop?
A. You should wait for a gap to be created.
B. You should be sure there is enough space.
C. You should not assume space will open up once you start pulling out.
D. All of the above.

23. How should you drive around curves?
A. Slowly
B. Fast
C. Gradually accelerating
D. None of the above

24. How should you watch out for trains at railroad crossings?
A. Listen.
B. Look.
C. Take a step out of your seat.
D. All of the above.

25. Your vehicle should be at least how many feet from a drawbridge?

A. 20

B. 30

C. 40

D. 50

16 School Bus Endorsement Test

1. Which of the following is an effective way of managing students?

A. Observe the school rules and regulations on discipline.

B. Keep driving until you arrive at your destination.

C. Do not turn off your engine.

D. Show that you are angry.

2. How do you speak to offenders?

A. Firmly but politely

B. Very seriously

C. Angrily

D. All of the above

3. Where should you remove an offender from the bus?

A. At the official bus stop

B. At the point of conflict

C. Wherever you decide

D. None of the above

4. When should you make a stop on a new route?

A. In an emergency

B. When the occasion demands

C. After requesting approval

D. When the route looks safer

5. Which of the following statements is true about arriving at a designated bus stop?

A. You should decrease the speed of your vehicle.

B. You should check for passersby.

C. You should not stop checking the mirrors.

D. All of the above.

6. How far ahead of the school bus stop should you switch on the warning lights?

A. 50 to 70 feet

B. 100 to 500 feet

C. 20 to 50 feet

D. 10 to 20 feet

7. Which indicator should you switch on when you are about to pull over?
A. Left
B. Front
C. Right
D. None of the above

8. How frequently should you check the mirrors?
A. Seldom
B. Once
C. Twice
D. Intermittently

9. When should you open the door?
A. Once you get to the bus stop.
B. After waiting for total clearance
C. When you need fresh air
D. All of the above

10. Where should students wait for a driver when going to school?
A. At home
B. Where the driver can see the students when approaching the bus stop
C. At the school
D. All of the above

11. Which of the following is true about loading students?
A. Learning the names of your students is helpful.
B. Taking a head count is not necessary.
C. You should not care if a student is missing.
D. All of the above

12. When loading the students, ask them to _________.
A. Use the handrails
B. Quickly go inside
C. Form two lines
D. All of the above

13. Which of the following should you do if students are crossing the road?
A. Step down from the bus and walk about 10 feet away to where you can see the students crossing.
B. Stay at the right side of the road where you can view the students' feet.
C. Look in all directions and ensure the road is clear for crossing.
D. All of the above.

14. Which of the following is included in the unloading guidelines?
A. Switch off the ignition.
B. Leave your key in the ignition.
C. Allow students to leave the bus as they like.
D. Stay on the bus.

15. After accounting for each student, what should you do?
A. Shut the door.
B. Put on your seat belt.
C. Run the transmission.
D. All of the above.

16. When should you leave the unloading base?
A. When there's still congestion
B. One hour after unloading
C. After traffic clears
D. All of the above

17. A danger zone may be as far as ________ feet from the front bumper.
A. 30
B. 90
C. 100
D. 150

18. You should ensure all the mirrors are _______.
A. Set facing the front
B. Set facing behind
C. Set giving you the right visual field
D. All of the above

19. A blind spot is ________
A. Where only your mirror can see
B. The side of the road that the mirrors cannot capture
C. 20 feet from the mirror
D. None of the above

20. The outside flat mirrors help you see approximately how many feet at the back of your bus?

A. 200

B. 500

C. 600

D. 800

21. Which of the following are/is found directly under the flat mirrors?

A. The outside flat mirrors

B. The outside convex mirrors

C. The outside crossover mirrors

D. The overhead rearview mirror

22. Properly adjusting the outside convex mirrors will help you have a view of ________.

A. Every side of the bus

B. The front part of the back tires where they have contact with the ground

C. The traffic lane outside

D. All of the above

23. Which of the following give/gives you a proper view of the front wheel and the service door areas?

A. The outside flat mirrors

B. The overhead rearview mirror

C. The outside convex mirrors

D. The outside crossover mirrors

24. Which of the following help/helps you check what is going on inside the bus?

A. The outside convex mirrors

B. The outside crossover mirrors

C. The outside flat mirrors

D. The overhead rearview mirror

25. A passive crossing means ________.

A. Stopping and starting when crossing the road

B. Applying normal traffic procedures

C. Not concentrating at crossings

D. None of the above

9 Answers for General Knowledge Test

(1) (B) Overbraking.

Overbraking is when you brake very hard or shut the wheels. It can result in skidding, which also happens when you apply the speed retarder on a slippery road. Skidding is when the tires do not have a grip on the road. Avoid overbraking by gently handling the brake.

(2) (D) All of the above.

Alcohol adversely affects the driver's judgment, time of reaction and vision. A drunk driver is not well coordinated and may run over the curb, weave, drive into a wrong lane and take longer to react to hazards. Alcohol causes a driver to commit errors that may lead to loss of lives and injuries. That's why it is illegal to drink before and during driving.

(3) (C) Turn on the parking brake and slowly pull against it.

To avoid danger, test your parking brake, emergency brake and service brake before driving. If you want to test the parking brake, stop the vehicle first. Then turn on the parking brake before pulling the gear low to see if the parking brake will stay.

(4) (C) Pour water on it.

When a tire catches fire, the best thing to do is to cool it down. You may have to use lots of water. You cannot remove the tire until you are sure it is cool. Then you can check for the cause of the fire.

(5) (B) 12 to 15.

Expert drivers look at least 12 to 15 seconds in front of them. This does not mean you do not see what is near you. It only proves that you are conscious of what is ahead and what is close. When not driving fast, 12 to 15 seconds equals one block. It is about a quarter mile at a higher speed.

(6) (B) Makes it difficult for drivers at the rear to have a clear sight of hazards in front.

There may be hazards ahead of you, and looking forward is the way to avoid them. These hazards are potential dangers for everyone. Drivers behind your vehicle may not be able to see ahead because of the size of your vehicle. Therefore, due to the size of your vehicle, if there is a hazard ahead, you may have to slow down and signal the drivers behind using your brake lights.

(7) (C) Not press the brake pedal.

When the spring brakes are on, leave the brake pedal. If you press the brake pedal, the brake may be destroyed due to the forces of air pressure and the spring. Many brakes, though not all, are designed this way.

(8) (D) All of the above.

As a cargo driver, you must be sure the weight of products is properly balanced. It is your duty to inspect your cargo before, during and after driving. Identify that your cargo does not restrict your access to emergency tools. Recognize if the loads are too much, then spread them out more evenly.

(9) (A) 1 block.

Expert drivers look at least 12 to 15 seconds in front of them. This does not mean you do not see what is near you. It only proves that you are conscious of what is ahead and what is close. When not driving fast, 12 to 15 seconds equals one block. It is about a quarter mile at a higher speed.

(10) (A) Adjustments.

Adjustment can affect the fading of a brake. Brakes that are not in adjustment will fail to work. The other set of brakes may then become too hot and fade out.

(11) (B) Check your mirrors.

Checking your mirrors helps you to know if there is anyone at your side or if someone is about to overtake you. Ensure there is sufficient room for your vehicle. Check again to ensure that no one is in your path before merging. Your mirrors are vital tools for checking the rear and sides of your vehicle to prevent accidents when changing lanes.

(12) (C) Avoid moving the steering wheel suddenly.

Trailers should be steered gently when being pulled. Moving the steering wheel suddenly can cause the trailer to tip over. Do not get too close to other vehicles. It is better to be as far as one second for every 10 feet of the length of your vehicle if the speed is below 40 mph and an additional one second if the speed is above 40 mph. Sudden lane changes should also be avoided.

(13) (D) All of the above.

You can prevent drowsiness by having enough rest and sleep before driving. It is better to find someone else to drive the vehicle if you are on medication that may lead to drowsiness. Scheduling the trip for the daytime is another option to consider.

(14) (A) After every trip.

The driver should review the trip report at the end of a trip and before another driver begins another journey. Every damaged component must be identified and repaired or replaced before the start of a new trip. This will help prevent problems that may pose a danger when on the road.

(15) (C) It is a factor that may lead to an accident.

When you feel like sleeping when driving, the best thing to do is to stop and take a nap. A nap of 20 minutes is better than 20 cups of coffee. Sleepiness may lead to an accident.

(16) (C) When the tires do not have a grip on the road.

A skid occurs when the tires do not have a grip on the road. Skidding occurs due to overbraking, oversteering, Over-accelerating or driving too fast. The newness or oldness of a tire has nothing to do with skidding.

(17) (C) Turn the steering wheel opposite the direction of the turn.

There is a difference between backing up a car or bus and backing up a trailer. When you are backing up a trailer, set the steering wheel in the direction opposite the turn. When backing up, stay in the right position, use your mirrors, check if your path is safe, back up slowly and use a helper.

(18) (D) When your vehicle is not in gear.

When you are approaching a turn, slow down, carefully change gears and coast safely. If your vehicle is not in gear, expect unsafe coasting. Your vehicle can be said to be out of gear if the gearshift is in a neutral position and the clutch is depressed.

(19) (C) The nuts are not well tightened.

When there is rust around the wheel, then it is likely that the nuts are loose. Checking the tightness of the nuts will help prevent unfortunate situations on the road. Drivers need to always take note of this before, during and after a trip. Therefore, it is necessary to stop after a while and check.

(20) (A) Slow down and set the transmission in a low gear.

You should not drive through flowing water or deep puddles. However, if it is necessary, slow down. Make sure the transmission is not in high gear. Hit the brakes slowly. Increase the RPM of the engine and slowly drive across the water without exerting much pressure on the brakes. Perform a test stop when you are sure it is safe to do so.

(21) (C) They make the stopping distance twice as long as normal.

It is not easy to stop on wet roads. It is very difficult to take a turn without skidding if the surface of the road is slippery. The stopping distance of wet roads is sometimes double that of dry roads. Decrease the speed by approximately one-third of the general stopping distance. Therefore, if the stopping distance is 75 mph, you should slow down to 50 mph.

(22) (C) Overaccelerating.

Overaccelerating is the supplying of too much force to the wheels and making them spin.

(23) (B) It may lead to skidding.

Most skids are from driving very fast. Therefore, good drivers need to adjust their speed.

(24) (C) Never release the clutch.

This is not a basic method for shifting up using manual transmission. It is not easy to shift gears when using double clutching. Do not try to force the gear. The vehicle may already be in neutral. Do not allow it to stay neutral for a long time. If it is difficult to switch to another gear, go to neutral, release the clutch and increase the speed of the engine to align with the road speed. Then try shifting the gear again.

(25) (A) A hazard.

When you see people working on any part of the road, you should consider it a hazard. A work zone may be a narrow lane, a sharp curve or an unbalanced surface. Keep your eyes on the road while driving to watch for hazards ahead. Do not get distracted. Drive gently when close to these zones. You should signal drivers with brake lights or four-way flashers as a form of warning.

(26) (A) Passengers in the vehicle.

Passengers in a vehicle are not a hazard. Hazards are people or objects that are obstructions to drivers. Hazards include pedestrians, bicyclists, road workers, ice cream trucks, disabled vehicles, crashes, children, talkers, distracted people, ramps, confused drivers, shoppers, slow drivers and drivers signaling a turn.

(27) (D) All of the above.

Holding on to the steering wheel tightly, staying on the brake and inspecting the front and rear tires are ways of responding effectively to tire failure.

(28) (B) Blood Alcohol Concentration.

BAC means blood alcohol concentration. It is the amount of alcohol in your blood or body system. The more you drink alcohol, the higher your BAC. When you drink very fast, it will increase your BAC. Your weight also affects how fast you reach your BAC. A small person does not necessarily have to drink as much to reach the legal limit of BAC as a bigger person does.

(29) (B) Tire failure.

You will know there is a tire failure by a loud sound, heaviness of the steering and vibration of the vehicle.

(30) (A) The rear tire is flat.

The thumping or vibration of a vehicle is a sign a tire is flat. You can know there is a tire failure by a loud sound, heaviness of the steering and vibration of the vehicle. Early identification of a tire failure gives you extra time to leverage. Holding on to the steering wheel tightly, staying on the brake and inspecting the front and rear tires are ways of responding effectively to tire failure.

(31) (D) The person's intellect.

A person's intellect has no impact on BAC.

(32) (B) Failure of a back tire.

Failure of a back tire will make the vehicle fishtail or go forward and backward. This is preventable with the use of dual tires.

(33) (D) All of the above.

An antilock brake system helps prevent your wheels from locking themselves. It is computerized and works with your normal brakes without reducing or increasing the effectiveness of your normal brakes. It is not a replacement for your normal brakes. Its sole function is to stop your wheels from closing up due to too much braking.

(34) (D) All of the above.

A bus, tractor or truck will have yellow ABS lights at the device component. There will be a red ABS light at the right side at the front or the back if it is a trailer, and the light will turn off when you are underway.

(35) (D) All of the above.

A high BAC affects your muscles, control, coordination and vision.

(36) (D) All of the above.

Before you drive, you must ascertain that you are healthy. Get enough rest and sleep. Do not drive if you are tired. Eat a balanced diet. Take time to visit your doctor regularly. Avoid medications that will make you sleepy.

(37) (D) All of the above.

When there is a minor fire while driving, call for help and make use of a fire extinguisher. Do not open the trailer doors because doing so may bring in air that will supply oxygen to the fire and increase the fire. If the fire is from hazmats, you may have to wait for experts to intervene.

(38) (D) Smoking and driving.

The following are some factors that cause vehicle fires:

- Fuel spill after crashes and wrong use of flares
- Touching duals and under-inflated tires
- Loose connections and damaged insulation, which makes the electrical system short circuit
- Smoking while driving and wrong fueling procedures
- Flammable cargo, wrongly sealed cargo and lack of sufficient ventilation.

(39) (B) Stay in the right lane.

Heavy-duty vehicles are usually tailgated if they find it difficult to align with the traffic speed. This often occurs when you are driving uphill and there is a heavy load pulling down your speed. Try to drive in the right lane. Do not drive past other slow vehicles when driving uphill unless you can do it safely and without wasting time.

(40) (A) It allows the lower maximum weight in axles that are not far apart.

When driving, you should keep the vehicle's weights around the lawful limits. Every state has a maximum axle weight. A bridge formula usually sets the maximum weight axle. This will stop overloading on bridges and roads. Overloading can adversely affect the control of speed, steering and braking when driving.

(41) (D) You will be allowed to drive vehicles that do not have air brakes.

An L restriction allows you to drive a vehicle that does not have air brakes if you did not take the skills test, if you did not pass the skills test and if the vehicle you used in the skills test does not have air brakes.

(42) (A) Stop and take a rest.

If you are feeling sleepy, take a break by safely parking and taking a nap or getting down to stretch your legs. Do not continue driving or trying to concentrate because you may fall asleep.

(43) (D) All of the above.

In the process of crossing or entering traffic, you should be aware of the size and weight of your vehicle because big vehicles accelerate slower and need more space, the rate of acceleration varies in every load and you want to avoid collisions with other vehicles. You need to be sure there is enough space for the size of your vehicle.

(44) (A) Stab braking.

Stab braking is the application of your brakes without stopping and letting go when the wheels are shut up. Resume the brake application when the wheels reopen and start running. After the brake release, the wheels may wait a second before they continue rolling. Wait until the rolling of the wheels begins before you press the brakes again. Otherwise, the truck may not be straight.

(45) (D) All of the above.

The engine needs to always be cool, especially in hot weather. You can do this through coolants such as antifreeze and water. Engine oil not only lubricates the oil; it also serves as a coolant that prevents overheating of the engine. Therefore, there should be sufficient engine oil in the engine.

(46) (B) Dynamite.

This is an example of an explosive. Other examples are ammunition and fireworks. They fall under Class 1 hazardous materials. They are vulnerable to fire.

(47) (C) Slippery surfaces.

Water or slush on the slippery surface of the road sometimes lead to hydroplaning. To regain control, let go of the accelerator and press the clutch. Do not use the brakes to reduce your speed.

(48) (D) None of the above.

You should not do any of these things. To ensure safety when proceeding through an intersection, carefully check traffic in every direction, reduce the speed, allow pedestrians to pass through, keep to one lane and have both hands on the steering wheel.

(49) (A) Slush.

Water or slush on the surface of the road sometimes leads to hydroplaning.

(50) (A) Black ice.

Black ice is a clear, thin layer of ice that is transparent enough for you to see the road below it. It gives the road a wet look. It appears whenever the temperature is below freezing. The road is slippery with black ice, so you need to drive carefully or find another route.

10 Answers for Air Brakes Test

(1) (A) Compressed pressure.

Air brakes are classified as one of the high-power or heavy brake systems. Air brake systems are often found in trucks, heavy vehicles and commercial buses. Using their legs, drivers apply great effort in operating this system. An air brake system will not work if compressed or accumulated pressure is not applied to the brake component.

(2) (D) It creates friction against the brakes and the tires.

The generated pressure from the compressed air uses the brake pad to create friction against the brakes and the tires. This is what makes the air brake systematically function. This system uses compressed air to supply greater force for the effective operation of the air brake system.

(3) (B) When driving.

The service brake is part of the three components of the air brake. It is meant for routine use when driving. It's not for emergency purposes, parking or pulling over. The service brake should be properly handled because if it is not maintained or utilized properly, it may result in heating or fading of the service brake.

(4) (D) When there is a failure of air supply to the brake chamber.

The hydraulic brake system is also known as the emergency brake. It is used when there is a brake system failure. It is very important when there is a failure of air supply to the brake chamber. It is mandatory that every bus, truck and other heavy vehicle have emergency brakes.

(5) (D) When compressed air makes the brake system functional.

The brake cylinder receives the compressed air traveling within the brake line system. The compressed air on the piston forces it to adjust from its position. The pneumatic energy, which was derived from compressed air, has been transformed into mechanical energy. This mechanism is what makes it easier to drive heavy vehicles.

(6) (D) The air compressor governor.

The air compressor governor of the air brake system controls when atmospheric air is supplied from the compressor to the air reservoir or storage tanks. It ensures there is proper timing of air supply. For instance, when the atmospheric air has reached a maximal level of 125 PSI, the air governor makes sure there is continual pumping of air from the compressor.

(7) (D) Below 3 PSI in 1 minute.

When testing for air leakage rate, first switch off the engine, let go of the parking brake and monitor the amount of time it takes for the air pressure to fall. Often, the air loss rate is below 2 PSI per minute if it is a single vehicle and below 3 PSI per minute if it is a combination vehicle. Proceed to exert 60 PSI and above using the brake pedal.

(8) (D) None of the above.

The storage tank makes it possible for the brake to be applied as often as necessary when the vehicle is on a trip. It serves as a reservoir for atmospheric air, which has been compressed under high pressure. It is vital because it sustains the movement of a vehicle on a trip when there is a lack of atmospheric air needed for the operation of the vehicle.

(9) (B) The brake pedal.

The brake pedal is an essential component of the brake system. The driver operates the brake pedal, which is inside the vehicle, manually. The brake pedal is a mechanical link that uses the applied effort on the mechanism for either slowing or stopping the movement of the vehicle. You need a brake pedal to stop suddenly.

(10) (C) It ejects air when the pressure gets to its maximum point.

The safety valve makes sure the tank does not burst when there is excessive pressure from the constant air that the compressor supplies. The safety valve frees air when the pressure gets to its maximum point of 150 PSI. It is like a regulator attached to the storage tank. It performs a supportive role to the storage tank.

(11) (B) Supply line.

Brake lines are also called supply lines. Their special function is transporting compressed or stored air from the storage tank to the brake drum. They stand in the gap between the storage tank and the brake drum. This is why they are also called supply lines; that is, they supply compressed air to the brake drum when the brake is applied.

(12) (D) It is a limitation to the back brake.

This statement is false. The front brake limiting valve is properly labeled as normal and slippery to enable better control of it. The pressure that travels to the front brake will be limited to half if you switch to slippery mode. A limiting valve inhibits the front wheel from slipping away when on slippery roads. If you put the control on normal, then it will be able to stop normally.

(13) (A) It removes water and oil residuals from the tank.

The air tank draining valve is meant to remove water and compressor oil residuals that may be in the tank. Failure to drain the water and oil may lead to faults in the air brake system if the liquids freeze out during cold weather. Draining of the air tank can be done manually or automatically.

(14) (B) Alcohol evaporator.

Most drivers do not know they have an alcohol evaporator in their air brake systems. The alcohol evaporator pours alcohol into the air brake system to avoid the consequences of trapped ice in the air brake system during cold weather.

(15) (B) The shaft of the brake cam.

The shaft of the brake cam operates when you apply the brake pedal. Applying the brake pedal allows air into the brake chamber, and the pressure from the air makes the rod protrude. This affects the slack adjuster, which twists the shaft of the brake cam and rolls the S-cam. However, when you let go of the brake pedals, the S-cam rolls back and brings back the brake shoes from the brake drum.

(16) (B) Brake chamber.
Disc brakes use an air mechanism in the application of air pressure against a brake chamber in a power screw slack adjuster, which resembles the S-cam. The pressure from the impact of the brake chamber against the slack adjuster moves the power screw. Then, the power screw hits the rotor, which is found on caliper brake lining pads.

(17) (B) The application pressure gauge.
The application pressure gauge shows you exactly the amount of air pressure you have made on the brakes. An application pressure gauge is not in all vehicles. The more pressure you apply to keep up with speed, the more this is a sign that your brakes may be fading and need immediate replacement.

(18) (D) All of the above.
The low air pressure signal tells you when your pressure supply is almost below 60 PSI or one half of your air compressor governor is running down on pressure. You will see a red light or hear a buzzer to notify you. Most big buses give off a warning when air pressure is at 80 to 85 PSI.

(19) (A) Your wheels will not lock.
An antilock brake system, which is also known as ABS, helps prevent your wheels from locking themselves. It is computerized and works with your normal brakes without reducing or increasing the effectiveness of your normal brakes.

(20) (D) All of the above.
The ABS helps prevent your wheels from locking themselves. When your ABS is malfunctioning, your normal brakes will still be available for use. However, you will have to quickly repair the ABS. Also, the ABS does not necessarily shorten your stopping distance, but it does help you keep the vehicle under control during hard braking.

(21) (A) Start by pulling down the brake pedals.
When you intend to stop your heavy vehicle normally with the ABS, pull down the brake pedals. Use a controlled pressure to arrive at a smooth stop. If the transmission is manual, wait until the engine RPM is idle before you push the clutch. Choose the starting gear after stopping.

(22) (A) A straight truck or combination vehicle that has ABS on both axles.
Do not change how you brake when driving a truck, tractor or bus with ABS. The only exception to this rule is when you are handling a straight truck or combination vehicle that has ABS on both axles. If there is an emergency stop, you can use the brakes without restraint.

(23) (B) Controlled.

The controlled braking technique keeps pressure hard on the brakes and will not lock the wheels. Do not make much movement on the steering wheel. If you must make a great steering adjustment of the wheel, then let go of the brakes and then go back on them again.

(24) (D) All of the above.

All these answers are false. Air brakes are meant for heavy vehicles. They are not only for road vehicles. Railways also use air brakes. With air brakes, high atmospheric pressure can be easily reserved in the storage tank. They make it easy to brake while driving heavy trucks and buses.

(25) (B) Check for tightness.

When inspecting the engine compartment, you will first have to look for the air compressor where the air governor is situated. You can find this where the driver sits. Check how tight the belt is if the air compressor uses a belt to function. Make sure the belt is in perfect shape. It should not be loose or show wear and tear.

11 Answers for Doubles and Triples Test

(1) (A) You use the same speed you use with combination vehicles to prevent a rollover.

This is not true. It is risky to transfer the knowledge for driving singles and combinations to doubles and triples. A reasonable speed for the former may be dangerous for the latter vehicles.

(2) (B) Long.

Doubles and triples occupy more space on the road than other vehicles do. They are longer.

(3) (D) All of the above.

Doubles and triples take a few seconds before they can be stopped completely. Therefore, do not get too close to other vehicles. Maintain a good distance. Do not be in a hurry to cross traffic. First, confirm if the space is long enough.

(4) (D) Slippery roads.

Slippery roads and fog often lead to road accidents. A double and triple vehicle requires greater skills in these adverse conditions. If the weather conditions are not ideal for driving, you should either wait for it to clear off or be extra careful on the road.

(5) (B) Difficulty in pulling out.

Careless parking may lead to difficulty when pulling out. Where and how you park your vehicle is very important with doubles and triples. Park where you can easily pull out. Observe the arrangements of the parking space to see if it fits your vehicle.

(6) (A) Engage the parking brake and press the red switch.

To ensure there is a maximum flow of air around the trailers, engage the tractor's parking brake and/or stop the moving vehicle by chocking the wheel. After waiting for the air pressure to become normal, press the red switch, which is meant to supply air to the emergency line. If you want the air to get to the service line, then apply the trailer's hand brake.

(7) (C) Air must go to all areas.

The air brake system needs an optimum amount of air around the vehicle for better functioning. Air must go to all the areas if you want the brakes to function effectively. You should hear the sound of air coming out from the air lines. If you do not, go back to the shut-off valves placed on the double or triple and the converter dolly. Then set it open.

(8) (A) Keep the trailer emergency brake switched off.

This is not a step when inspecting emergency brakes. When inspecting the trailer's emergency brakes, power up the air brake system and ensure there is no obstruction to the rolling of the trailer. Then stop and bring out the air supply control. Or you can simply set it in an emergency position. Slowly pull out the trailer and the tractor to ensure that the trailer emergency is switched on.

(9) (D) Use the wheel to inspect the trailer's brakes.

This is not a step when inspecting the service brakes. When inspecting the service brakes, keep the air pressure normal. Gently drive the vehicle to the front. Press the trailer's brakes using the trolley valve or hand control. Use the hand valve to inspect the trailer's brakes, but use your foot pedal if it is for a basic function.

(10) (A) Coupling.

A converter dolly is a tool for coupling one or two axles together, including a fifth wheel. You can couple the semitrailer to the back of a tractor-trailer. The second trailer should be behind the converter dolly. Your vehicle needs proper coupling to prevent disjointing when driving.

(11) (D) All of the above.

The rear trailer may not possess spring brakes. If that is the situation, bring the tractor and the trailer closer, integrate the emergency line, charge up the trailer's air storage tank, then set the emergency line apart. This setting will put up the emergency brakes after correctly adjusting the slack adjusters. If you doubt the functionality of the brake, chock the wheels.

(12) (D) All of the above.

When using the front semitrailer and the tractor to pick up the converter dolly, set the combination very close to the converter dolly, lock the pintle hook, protect the dolly in a lifted position, bring the nose of the rear semitrailer closer to the dolly, bring down the dolly support and remove the dolly from the trailer that is right behind the tractor.

(13) (C) Connect the emergency line and the service line together.

This is false. When coupling the converter dolly with the front trailer, put the first semitrailer in front of the dolly tongue. Connect the front trailer with the dolly. Lock the pintle hook and safely set the converter gear in a lifted position.

(14) (A) A bit below the middle of the fifth wheel.

When coupling the converter dolly with the back trailer, you will have to set up the height of the trailer. The trailer should be a little bit below the middle of the fifth wheel so that the trailer can go a bit higher when the dolly receives a thrust from below.

(15) (A) The brakes will not work if there is no air supply and storage.

Air brake systems require air to function. Without the supply and storage of air, the brakes will not work because these brakes need the pressure from compressed air to function. The size of air brake systems makes it difficult to use physical force only.

(16) (B) Lock up the second trailer wheels.

If your vehicle does not possess spring brakes, lock up the second trailer wheels when uncoupling the back trailer of your twin or double trailers. Other steps to uncoupling the back trailer include setting the rig firmly on clear ground and within a strategic sequence and pressing the parking brake so that the rig will be stabilized, among others.

(17) (B) Landing gear.
Releasing the landing gear of the rear semitrailer will take away the weight or pressure from the dolly. When uncoupling twin or double trailers, set the rig firmly on clear ground and within a strategy sequence. Press the parking brake. If your vehicle does not possess spring brakes, lock up the second trailer wheels. Lock up the shut-off behind the front trailer.

(18) (A) They come with spring brakes.
Converter dollies often come with spring brakes. When uncoupling a converter dolly, bring down the dolly's landing gear. Remove the safety chains. Press the converter's spring brakes or chock the wheels. Unlatch the pintle hook, which is on the front semitrailer. Gently drive away from the dolly.

(19) (B) Unlatch.
When uncoupling converter dollies, you should unlatch the pintle hook, which is on the front semitrailer.

(20) (B) Couple.
If the dolly is beneath the back trailer, do not open the hook of the pintle, or the dolly bar will fling up. This may result in injury and make it difficult to couple again.

21. (B) Pull out the dolly.
When you want to uncouple a triple-trailer rig, pull out the dolly to disjoint the third trailer from the other trailers and tractor. Other steps to coupling and uncoupling of trailers include applying the process for joining tractor to semitrailers and setting the converter dolly in its place, then applying the procedures for joining doubles in coupling the first and second trailers together.

22. (D) All of the above.
When inspecting doubles and triples, make sure there is no fault in any of the components. Ensure the glide plate is safely placed on the frame. Ensure the kingpin does not get faulty. Ensure there is no fault in any of the air lines. Ensure the fifth wheel slide is not faulty or missing any of its components.

23. (A) Grease.
Grease serves as the lubricant for all areas of your doubles and triples, especially the engine. Your vehicle should be well greased at all times.

(24) (A) Only fix the emergency air line.
This is not a step in inspecting the components of doubles and triples. You should keep the valves closed. Ensure all air lines are well fixed. Secure the spare tire placed on the converter gear. Ensure

proper locking of the pintle. Provide security for the safety chains. And properly connect light cords to the sockets on the trailers.

(25) (A) Locked.
The pintles of doubles and triples should be locked during an inspection.

12 Answers for Hazardous Materials (Hazmats) Test

(1) (C) They are risky only to the health of the driver.

This statement is not true. *Hazmats* is an acronym for *hazardous materials*. They are potential dangers that may have adverse effects on the safety, health and assets of anyone in the course of transportation.

(2) (A) Compulsory for every driver transporting hazardous materials.

If you want to transport hazardous materials, you must obtain a commercial driver's license (CDL). Before you are given the license, you must pass a test that requires you to know the regulations of transporting hazardous materials. Employees and drivers handling hazardous materials must go through special training once every three years.

(3) (B) Class 2.

Class 2 hazardous materials are compressed gases, such as helium and propane. Warn everyone around if you notice any leaks. When there is a crash, allow only those who engage in unloading the hazardous materials in the environment. Contact the shipper immediately. Warn others to keep away when there is an accident. It is dangerous to transfer gases from one tank to another in public places.

(4) (C) Class 5.

Class 5 hazardous materials, such as hydrogen peroxide and ammonium nitrate, are also known as oxidizing materials. Just like Class 4 materials, they are potential dangers to the safety of everyone around.

(5) (B) Plutonium and uranium.

Plutonium and uranium are examples of radioactive substances. They fall under Class 7 of hazardous materials. Radioactive substances are potential dangers to the safety of everyone around. Do not allow smoke or fire to come close to your vehicle if there are radioactive materials inside. Warn others to keep away when there is an accident.

(6) (B) Containment rules.

The hazmat regulations are also known as containment rules. They instruct drivers and companies on the safest ways for packaging, loading and unloading hazardous materials. They are meant to prevent the explosion and exposure of hazardous materials to a harsh environment. These regulations are not punishments for offenders. Instead, they are meant to protect everyone.

(7) (D) All of the above.

When you try to use shortcuts and break hazmat rules, the result is imprisonment. Therefore, it is better to master and apply these regulations for your safety. Law enforcement officers are on the road to ensure the rules of hazmat are applied and to check if you have the necessary papers. Failure to apply hazmat regulations may lead to accidents and death.

(8) (A) 2.

The size of the shipping name should not be less than two inches in height if the capacity of the portable tank is above 1,000 gallons and one inch tall if the capacity is below 1,000 gallons.

(9) (B) Intermediate bulk container.

The intermediate bulk container is a type of bulk package that does not need a shipping name. The owner's name does not have to be on it. You can proceed to load hazardous materials into it without name markings on it.

(10) (A) Static.

Do not load or unload when you have not adjusted the parking brake. Ensure the vehicle is static. This is to prevent accidents while loading and unloading.

(11) (A) Packed tightly.

The following items should be packed tightly to avoid movement or sliding of the products when they are being transported: gases, oxidizers, flammable liquids and solids, explosives, corrosives, radioactive and poisons. Safely brace containers that house these containers. Two or more substances may hit each other or fall and break open when the vehicle hits a bump or goes around a curve.

(12) (B) Prohibited.

Anything that produces fire or heat should not be brought close to hazardous materials during loading and unloading. Keep materials far from heat sources because most hazardous materials are flammable. Heat is a potential danger to the lives of everyone and the assets nearby. Therefore, smoking is prohibited around such materials.

(13) (B) Is unloaded and loaded before it is placed on the vehicle.

A bulk packaging that is not fixed on a vehicle is called a portable tank. The portable tank is not on the vehicle during loading and unloading. Every portable tank should display the owner or lessee's name.

(14) (C) MC331.

There are different types of cargo tanks. The most popular one is the MC306, which is meant for hazardous liquids, and the MC331, which supports gases.

(15) (D) All of the above.

The shipper is one of the key stakeholders in the transportation of goods. The shipper is in charge of transporting them to different points; applying hazmat regulations in identifying a product; classifying hazards; providing identification numbers; grouping for packing; correctly packaging products; labeling, marking and placing placards; and making sure all the shipping papers have been obtained.

(16) (A) A tool for communication.

When there is an accident, you may be unable to speak and find it difficult to communicate the types of hazardous materials that are in your vehicle. The police and first responders can minimize the level of risk if they are aware of the kinds of hazardous materials that are in the vehicle. The solution to this is the contents of the shipping papers.

(17) (D) All of the above.

There are different places where the driver can keep the shipping papers. They can be in a bag on the driver's door or the driver's seat or within arm's reach of the driver. What matters is that they can be easily found in an emergency.

(18) (A) Placards.

Placards signal other drivers of the presence of hazardous materials.

(19) (B) Outside.

Placards are signs situated outside a truck or tanker and on top of bulk packages that contain hazardous materials.

(20) (C) Four.

A placarded vehicle requires a minimum of four clearly labeled signs.

(21) (C) On the front, back and sides of the truck.

Placards are placed in the front, back and sides of the truck.

(22) (A) They should not be close to explosives, blasting agents, oxidizers or poisonous gases.

When loading corrosive liquids, do not load them close to explosives, flammable solids, oxidizers and poisonous gases.

(23) (A) Provide proper ventilation.

Proper ventilation is necessary to avoid overheating or the explosion of hazardous material.

(24) (D) All of the above.

Racks should be used for storing cylinders. If there are no racks, then ensure the floor of the cargo is flat. You may keep the cylinders right-side-up or in a horizontal position. Any package that is labeled as poison should not be kept in the driver's cab or with food materials.

(25) (D) All of the above.

A cargo tank is a type of bulk packaging. It contains tools, liquids, gases and reinforcements. It is a permanent or nonpermanent appendage to a vehicle. Due to its attachment to the vehicle, it can be loaded and unloaded without being detached from the main vehicle. It is not specifically designed for cylinders or any type of tanker.

13 Answers for Combination Vehicles Test

(1) (A) Its stability.
Piling too much cargo into a combination vehicle will affect its stability. The truck will lose balance and can easily tilt off the road. A loaded truck is more likely to roll over than an empty truck.

(2) (C) One second; ten feet.
The rule of keeping a safe distance specifies at least one second for every 10 feet of vehicle's length at a speed under 40 mph. When the speed goes up, increase the time by one second to enable safety. If you are driving a 20-foot vehicle, then the distance of time is a two-second gap. However, if the speed is over 40 mph, have a five-second distance for a 40-foot vehicle.

(3) (C) There is an increase in the following distance.
When you have an increase in following distance, it is very easy to apply your brakes when necessary, especially in emergencies. Driving too fast can make it difficult to apply your brakes. It is advisable not to drive too close to the vehicles in front.

(4) (B) When the front and rear wheels are in different positions when you are going around a corner.
The front and rear wheels are often in different positions when you are going around a corner. This is called cheating or off-tracking. Expect longer trailers to have more cheating than shorter ones. To come back in alignment, steer the front very wide around the corner so that the back end does not go up on the curb or hit pedestrians or anything else in the way.

(5) (A) The trailer hand valve.
Some people prefer to call the trolley valve a trailer hand valve or a Johnson bar.

(6) (D) All of the above.
The Johnson bar, the trailer hand valve and the trolley valve are used to test the brakes and should not be engaged in driving because the vehicle may skid.

(7) (B) Air leaks.
The trailer hand valve should not be involved when parking because it might lead to leakage of air. Instead, use parking brakes to park your vehicle and wheel chocks to prevent movement. The trailer hand valve is used to test the brakes and should not be engaged in driving because it may later skid.

(8) (B) It will prevent air from escaping.
The tractor protection control valve maintains the air within the air brake system if there are leaks in the trailer or if it breaks. If the air pressure is around 20 to 45 PSI, then the tractor protection valve will close up without manual application. This closure prevents air from escaping the air brake system. This mechanism makes the air go through the emergency line.

(9) (B) It has a red eight-sided knob.

You can identify the air supply control valve by its red eight-sided knob found in modern vehicles. The air supply control is used to supply and shut off the air. When you push it, there is an air supply, and when you pull it out, the air supply is shut off.

(10) (C) The service air line.

The service air line is also called a signal or control line. It transfers air around the air brake system. The service line pressure is based on the effort you exert on the brake or hand valve. The service line is linked to the relay valves, which facilitate the quicker application of trailer brakes.

(11) (B) Supply line.

The emergency air line is also called a supply line, and it has two basic functions. The first purpose is to supply air to the air tank, and the second is to control the emergency brakes.

(12) (C) Blue.

The service air line is also called a signal or control line, and it is blue.

(13) (D) Dummy couplers

Some combination vehicles are designed with dead ends, which are also called dummy couplers. They are connected with the hose when they are idle. This keeps water and dirt particles away from the couplers and the two air lines. Dummy couplers should be used when the emergency and service lines are not connected with the trailer. In the absence of dummy couplers, you can lock both glad hands.

(14) (C) They allow air line closure when there is no towing trailer.

Shut-off valves, or cut-out cocks, are for the supply or emergency line and the service air line, which are located behind towing trailers. These valves allow the closure of the air lines when there is no towing trailer. This is to prevent air leaks in the vehicle.

(15) (D) All of the above.

In coupling a tractor-semitrailer, you should observe the area. See if the vehicle area is clear. Ensure the spring brake is on and the wheel is properly chocked. Make sure the cargo is intact and not susceptible to movement. These measures are to avoid accidents and disturbances while coupling the tractor.

(16) (B) Coupling may become difficult.

The trailer should be reasonably low so that the tractor can bring the trailer up when it is backed below the trailer. A very low trailer may make the tractor hit the trailer and destroy the trailer's nose. A very high trailer may make coupling difficult.

(17) (A) Emergency.

While coupling the air lines to the trailer, make sure the glad hands are intact, then link the emergency line to the trailer's emergency glad hand. Connect the service air line to the trailer's service glad hand. Ensure the protection of air lines so that they are not crushed when the tractor is taking the back of the trailer.

(18) (A) Stop immediately.

When backing a tractor below a trailer, you should apply the reverse gear, which is at the lowest position. Go slow when backing up the tractor under a trailer and stop immediately once there is an interlock between the fifth wheel and the kingpin.

(19) (C) You should lift the gear and stop midway.

This is not true. When lifting the landing gear, start from the lowest point. When the landing gear has become weightless, increase the gear range. Keep on lifting the landing gear until it is at the top and not halfway to avoid it being trapped on railroad tracks.

(20) (D) None of the above.

When coupling tractors, you should set the rig in the right position, reduce the pressure against the locking jaw, chock the wheels to prevent movement and put down the landing gear.

(21) (D) To lock the brakes.

When reducing the pressure against the locking jaw, first switch off the air supply to lock the brakes, then back up slowly to reduce the pressure on the fifth wheel. Turn on the trailer's parking brakes when the tractor is impacting the kingpin.

(22) (A) Lifted.

The fifth wheel should be left lifted so that it will not be difficult to unlatch it and coupling will still be easy. When putting down the landing gear, go ahead and bring down the landing gear until it touches the ground. After you have made the landing gear touch the ground and the trailer has been loaded, turn the crank slowly to cut off the extra weight of the tractor.

(23) (C) Make the recoupling easier.

The fifth wheel should be left lifted so that it will not be difficult to unlatch it and recoupling will still be easy. When putting down the landing gear, go ahead and bring down the landing gear until it touches the ground. After you have made the landing gear touch the ground and the trailer has been loaded, turn the crank slowly to cut off the extra weight of the tractor.

(24) (D) Place the release handle lock in an open position.

When uncoupling the fifth wheel, place the release handle lock in the open position. Stand far from the back of the tractor wheels to prevent injury that may occur if the vehicle moves. Be conscious of the fact that proper care is needed when coupling and uncoupling trailers.

(25) (B) Prevent the trailer from rolling.

When testing the trailer emergency brakes, ensure the rolling of the trailer is not prevented. Power up the air brake system and be sure the trailer can roll. Put the air supply control on emergency or pull out the tractor protection valve. Then gently pull the trailer and the tractor to be sure that the emergency brakes are on.

14 Answers for Tanker License Test

(1) (C) 119 and more.

You will need a tanker license if your tanker is transporting either gases or liquids in a fixed cargo. The cargo tank capacity may be 119 gallons or more, or it can even be a portable tank that has a capacity of 1,000 gallons or more. Every vehicle that requires a Class A or Class B CDL needs a tanker license.

(2) (C) Inspect.

No matter how solid your vehicle may seem, do not load, unload or drive until you have inspected its condition. You perform this inspection to ensure the safety of the liquid or gaseous contents and to ensure the safe driving of your vehicle to its final destination.

(3) (B) Liquid turbulence.

Tanker vehicles are unlike other vehicles. You need special knowledge and skills to drive one. This is because there is great liquid turbulence and an accentuated center of gravity involved. Therefore, you should not apply to tankers the same method of driving you use with other vehicles.

(4) (A) The vehicle can easily go off the road.

A greater center of gravity means that a greater part of the weight of the contents is raised high and almost off the road. Therefore, the top of the vehicle becomes weighty and is likely to fall over. It is easier for liquid loads to roll down. Rollovers are most likely to occur in curves.

(5) (A) Liquid surge.

When the tank is nearly full, expect liquid movement to lead to a liquid surge. The liquid movement will negatively affect your control of the vehicle. When you are stopping, the liquid will move forward and backward.

(6) (A) The wave might move the tanker in its direction.

When a tanker is parked on slippery ground, the wave can move the vehicle in its direction.

(7) (B) Partitioning liquid tanks.

Bulkheads are used to partition liquid tanks into smaller units. Any time you are loading or unloading, make sure that there is a balanced spreading of the weight. Avoid overloading the front or the back with excess weight. When you use bulkheads, there will not be an excess load on one part of the vehicle.

(8) (C) Baffled tanks.

Liquid tanks with bulkheads are also known as baffled tanks. They have holes that allow the passage of liquid. You can have better control of the front, back or sideways surge of liquid when you use baffles.

(9) (D) All of the above.

Baffled tanks are liquid tanks with bulkheads. They have holes that allow the passage of liquid. You can have better control of the front, back or sideways surge of liquid when you use baffles. Remember, liquid surge should be avoided as much as possible because it causes the content to run over.

(10) (A) Smoothbores.

Unbaffled tankers are also known as smoothbores. Unbaffled tankers do not have any tool to reduce the surging of the liquid. If you use unbaffled tankers to transport liquid content, expect a very high movement of the tanker. Unbaffled tankers are better used for transporting food products like milk.

(11) (A) Room for expansion.

An outage is the space left to make room for a liquid to expand.

(12) (D) All of the above.

Avoid loading a tanker to its brim. This is because liquid expands when warm, and this new volume needs space. Hence the need for an outage. All liquids do not expand at the same rate. Therefore, you need to be cautious of the type of liquid and the outage you are leaving. Know the outage rate before you load the liquid.

(13) (D) All of the above.

You will know the liquid is dense by how much it expands during transportation, its weight and if it meets the legal weight limit. Therefore, if the liquid is very dense (acid), a completely full tank may be beyond the legal weight limit. So, do not totally fill up the tank.

(14) (D) All of the above.

You have to be very careful when driving tankers. You need to drive smoothly at all times. This is because of the liquid surge and high center of gravity. Slow down, change lanes and stop smoothly. Do not go around curves too fast. You need extra care to avoid turbulence.

(15) (A) Decreasing stopping distance.

This is false. To take control of liquid surge, give the brakes consistent pressure. Take it slow when you are stopping. Maintain a safe following distance between you and the vehicle in front. Increase the stopping distance if the road is wet. Always apply the brake ahead of the stopping area. Apply stab or controlled braking to stop quickly and prevent a crash.

(16) (B) Make it tip over.

Driving your tanker too fast could cause it to tip over.

(17) (C) Skidding.

Too much acceleration or overbraking may make the vehicle skid. Skidding may make your tanker jackknife.

(18) (D) All of the above.

No matter how solid your tanker vehicle may seem, do not load, unload or drive until you have inspected the condition of the vehicle. Inspection is key to the prevention of accidents.

(19) (A) Drivers should know how.

This is not true. Walk around and check for leaks beneath or in any part of the vehicle. It is very risky to load and drive a vehicle with leaks. It is also illegal. If you are caught, authorities will stop you from continuing with the drive. And you may be mandated to clean up any spill along the route you have been driving.

(20) (A) Free from particles.

To be thorough with your inspection for leaks, check the body or outer part of the tank, the valves, all the connections and pipes, the vents and manhole covers. Ensure there are available gaskets that keep the covers well closed. Ensure the vents are free from particles or obstructive substances for efficient functioning.

(21) (A) It is always labeled red.

This is not true. Special purpose equipment is not always labeled red. It includes the grounding wires, the bonding wires, the vapor recovery package and the fire extinguisher that comes with it.

(22) (D) Semi-inspection.

This is not a type of vehicle inspection for a tanker. The pre-trip inspection is carried out before driving. It helps you prevent a breakdown on the road. During the trip, an inspection is carried out while driving. It includes checking the lights, the tires, the cargo, etc., at intervals. An inspection is also carried out after the trip.

(23) (B) Prevents your wheels from locking themselves.

An antilock brake system, which is also known as ABS, helps prevent your wheels from locking themselves.

(24) (A) Safely reduce the speed of runaway vehicles on a steep mountain.

Escape ramps are made to reduce the speed of runaway vehicles on a steep mountain downgrades. Without an escape ramp, there would be a loss of lives, equipment and cargo. Learn to identify the locations of these escape ramps.

(25) (D) All of the above.

Safety is the priority of every driver. Before, during and after driving, you are to carry out a safety inspection to avoid accidents.

15 Answers for Passenger Transport Test

(1) (B) Banned from your vehicle.

Riders sometimes board buses with unlabeled hazardous materials that are potential dangers to the lives of other passengers. Therefore, do not allow riders to carry common hazmats such as car batteries or gasoline into your vehicle. Do a proper inspection, because some of these hazmats might implicate you and everyone else if the police discover the existence of these materials in your vehicle.

(2) (D) All of the above.

You must ensure every object and person you are driving is safe before loading and embarking on a trip. Go through the inspection report that the previous driver made to see what is lacking. Inspect the parts of the vehicle and confirm that they are all highly functional.

(3) (D) None of the above.

All of these are false. You should go through the inspection report that the previous driver made to see what is lacking. Refuse to sign it if what needs to be repaired or replaced has not been effected. Do not sign it unless it is reported that the damaged component does not require repairs and you are certain of that.

(4) (D) All of the above.

Before you drive, inspect the following components in order: the parking brake; the service brakes, which may include air hose couplings for vehicles with a semitrailer or trailer; the reflectors and the lights; the steering system; the horn; the tires; the wipers; the mirrors; coupling tools; the wheels and the emergency kits.

(5) (D) All of the above.

Before you load and drive your vehicle, check the interior and ascertain that it is safe for everyone to go in. Check if the seats are safe for passengers and the driver. Inspect the handholds and the railing, the covering of the floor, the signal tools and the handles of the emergency exit.

(6) (A) Closed.

The emergency window is an interior part of the vehicle. Keep the emergency window and door closed when driving.

(7) (B) Boldly.

Make sure the inscription of "emergency exit" is very boldly and clearly written on the emergency door. This will be a guide to every passenger so that they will know which door to exit from in an emergency. The visibility will enable those with poor sight to see the inscription. Also, always keep the emergency light on at night.

(8) (D) All of the above.

Ensure the fire extinguisher is available, functional and filled.

(9) (D) Drivers and passengers.

Seat belts are essential components of the vehicle. They are meant to prevent passengers from jolting out of their seats. Inspect the seat belts in the vehicle. Always make use of yours and advise your passengers to use theirs. If there is wear and tear on the seat belts, replace them with new and stronger ones.

(10) (B) Be forbidden.

Keeping any baggage by the door of a bus is strictly forbidden for safety reasons.

(11) (B) A possibility in passenger buses.

Smooth movement within a vehicle is a possibility if appropriate procedures are followed.

(12) (D) All of the above.

You are not allowed to transport certain hazardous materials on passenger vehicles. These include poison gas, irritating material, or liquid poison. You are also not allowed to transport solid poisons that weigh more than 100 pounds.

(13) (C) ORM-D.

Passenger buses are allowed to drive small-arms ammunition that is ORM-D, emergency supplies for hospitals and drugs.

(14) (B) They should not be kept near passengers' seats.

You are not allowed to keep radioactive materials and explosives near passenger seats.

(15) (A) They should not be allowed.

You are not allowed to transport certain hazardous materials on passenger vehicles. This includes gasoline and battery cells, which may explode under certain conditions and result in fatalities.

(16) (A) 2.

Vehicles that are made to allow standing should have a marked standee line of two inches. The standee line shows passengers where they can and cannot stand.

(17) (C) Announce the time you are stopping.

When you are almost at a bus stop or final destination, you should announce the destination, state why you are about to stop, announce the time of departure and announce the number of the vehicle.

(18) (B) Keep everyone outside until you are ready to move.

To avoid stealing and vandalism, keep all passengers outside a vehicle until you are ready to move.

(19) (A) Continues while on the road.

Your responsibility as a driver continues while on the road. It does not end at the beginning of the trip, and neither is it just for the end.

(20) (D) All of the above.

When you are at a bus stop, you have to be vigilant. Some passengers may fall off when entering or exiting the bus. Always remind them to be cautious of their steps to avoid stumbling. And before you start driving, wait for every rider to sit comfortably to avoid injury.

(21) (C) Intersections.

Most bus crashes occur at intersections. Do not depend on traffic signs alone; use discretion when driving passenger vehicles. Be observant enough to notice incoming vehicles, poles, or objects in the way. Sometimes you may not be the driver at fault, but you still need to prepare for reckless drivers.

(22) (D) All of the above.

Be sure there is an available gap before you join a lane or enter traffic after a stop. Do not expect that other drivers will give you space to fill in after you have started pulling out. Wait for the gap to be created before you fill in the space.

(23) (A) Slowly.

Slow driving is far better than fast around curves. Decelerate for everyone's safety. Driving fast often results in the loss of lives and property.

(24) (D) All of the above.

When stopping at railroad-highway crossings, apply the brake when your bus is about 15 to 50 feet from the railroad crossings. Watch out for trains by listening and looking. If possible, take a step out of your seat to be sure if there are incoming trains or not.

(25) (D) 50.

If the drawbridge does not have a traffic control attendant or green signal light, apply your brakes. It is not compulsory to stop, but you have to slow down at least 50 feet from the drawbridge. Inspect to ensure the drawbridge is fully closed up.

16 Answers for School Bus Endorsement Test

(1) (A) Observe the school rules and regulations on discipline.
When managing students, observe the school rules and regulations on discipline before, during and after transit.

(2) (A) Firmly but politely.
When there is a dispute, speak firmly but politely to the offender(s). Do not express anger in your tone or facial expression, but be serious so that those involved will not make the situation worse. Do this after you have stopped the bus in a safe parking area.

(3) (A) At the official bus stop.
On no occasion should you remove a student from the bus except when you have arrived at the official bus stop. You may call the school administration or the police and inform them so that they can come and pick up the student if the offense is very serious. Just remember to comply with protocols on discipline.

(4) (C) After requesting approval.
Schools have approved routes and bus stops. If you notice a new route or bus stop that should be used, do not use the new route or make a stop until you have officially written to the school district for approval. This is to ensure the safety of the students.

(5) (D) All of the above.
When arriving at a designated bus stop, decrease the speed of your vehicle and check for passersby, objects and traffic as you are stopping. Even after you have stopped, check again. Switch on the warning lights a few seconds (5 to 10) or 100 to 500 feet before you arrive at the stop.

(6) (B) 100 to 500 feet.
Before you arrive at the school bus stop, switch on the warning lights a few seconds (5 to 10) or 100 to 500 feet ahead.

(7) (C) Right.
Do not just pull over without switching on your indicator. Switch on the right-side indicator three to five seconds before you pull over. This will help prevent collisions with other vehicles.

(8) (D) Intermittently.
When arriving at a designated destination, intermittently check the mirrors. You need to be sure of everything that is going on around you and be vigilant. Do not set the transmission in neutral or park. Wait for total clearance. You can open the service door to switch on the red lights if the traffic is a bit far from the school bus.

(9) (B) After waiting for total clearance.

Wait for the traffic to be totally clear before you open the door and ask the students to start coming in. You do this to prevent collisions with any other driver who may be passing by. Stay very close to the side of the driving lane to prevent accidents.

(10) (B) Where the driver can see the students when approaching the bus stop.
Students are instructed to wait for the arrival of the school bus at a specific bus stop. This location should be where the driver can see the students when approaching the bus stop.

(11) (A) Learning the names of your students is helpful.
Always remember to take a count of the students and make sure none misses boarding the bus. You can learn the names of the students. If the number of students on board does not tally with the expected number, ask for the whereabouts of the missing student(s).

(12) (A) Use the handrails.
When the students are entering the bus, ask them not to rush in. Instead, they should be in a single-file line and use the handrails. Turn on the dome light if the loading area is dark. Avoid being hasty when loading. Ensure everyone is seated and ready to be transported before embarking on the trip.

(13) (D) All of the above.
If the student(s) will cross the road, step down from the bus and walk about 10 feet from the bus to where you can see the students crossing. Stay at the right side of the road where you can view the feet of the students. Look in all directions and ensure the road is clear for crossing.

(14) (A) Switch off the ignition.
When unloading students at school, switch off the ignition. Take your key with you when leaving your seat. Ask the students to keep sitting down until they are told otherwise. Go where you can oversee the unloading. Ask the students to step down in an organized pattern, row by row.

(15) (D) All of the above.
After you have accounted for each student, prepare to leave by shutting the door, putting on your seat belt, turning on the engine, running the transmission, letting go of the parking brake, switching off the alternating flashers, switching on the left turn signal and looking at the mirrors again. Then drive away.

(16) (C) After traffic clears.
You can leave the unloading base after ensuring that the traffic has cleared.

(17) (A) 30.

The danger zone of a school bus is an area where the schoolchildren are likely to receive the greatest injury from their bus or another bus. The danger zones of your bus may be 30 feet away from the front bumper, although the first 10 feet are often more dangerous.

(18) (C) Set giving you the right visual field.

To avoid accidents when driving, you need to properly adjust the mirrors, which are very useful for looking out for traffic, objects and the welfare of the students. Ensure all the mirrors give you the right visual field.

(19) (B) The side of the road that the mirrors cannot capture.

The blind spot is the side of the road that the mirrors cannot capture. A blind spot is directly in front of and under each flat mirror, as well as the back of the rear bumper. The distance of the blind spot is about 50 to 150 feet; sometimes it goes as far as 400 feet based on the size of the school bus.

(20) (A) 200.

When you properly adjust the outside flat mirrors, it will help you have a view of approximately 200 feet or the length of four buses behind your bus, the sides of the bus and the back tires where they make contact with the ground.

(21) (B) The outside convex mirrors.

The outside convex mirrors are directly under the flat mirrors outside your vehicle. You use them to have a panoramic view of what is happening at the left and right sides of your environment outside. Like the flat mirrors, they give you a view of the students at the side, traffic and other activities. These mirrors will not give you the exact distance and size of objects.

(22) (D) All of the above.

Properly adjusting the outside convex mirrors will help you have a view of every side of the bus, the front part of the back tires where they make contact with the ground and the traffic lane outside.

(23) (D) The outside crossover mirrors.

The outside crossover mirrors are on the left and right front sides of your vehicle. They give you a proper view of the front wheel and the service door areas. With these mirrors, you will have a good view of danger zones in front of and outside the bus. These zones are difficult to see with the eyes alone.

(24) (D) The overhead rearview mirror.

The overhead rearview mirror is suspended above the windshield in the driver's area. Instead of turning your head around and taking your concentration totally off the road, you should use the overhead rearview mirror to monitor activities inside the bus.

(25) (B) Applying normal traffic procedures.

Passive crossings do not have a traffic control machine. You have to stop and apply normal traffic procedures. Identify if there are imminent trains coming before you cross. Passive crossings come with round yellow warning signs and markings for easy recognition. Be vigilant when you are at passive crossings.

17 General Knowledge Test

1. Hitting objects in front or overhead is ______________.
A. Sometimes a necessity
B. A danger
C. A way to make it easier to move ahead
D. None of the above

2. The space below a vehicle is ____________.
A. Always the same in loaded and unloaded vehicles
B. Very small when the vehicle is fully loaded
C. Very big when the vehicle is fully loaded
D. Very small when the vehicle is unloaded

3. Which of the following prevents a crash when turning right?
A. Turning fast before other vehicles catch up with you
B. Avoiding turning wide
C. Setting the back of your vehicle near the curb
D. Turning wide to the left when you start turning

4. Antilock brakes ____________.
A. Help you closely follow drivers in front of you
B. Decrease the distance for stopping
C. Are an alternative when there is poor brake maintenance
D. Could help avoid a crash

5. Adding together perception distance, reaction distance and braking distance equals _______.
A. Total stopping distance
B. Average speed
C. Time taken
D. Time of departure

6. The service line is also called the ___________.
A. Cable line
B. Control line
C. Server line
D. Emergency line

7. What is the main reason for performing a vehicle inspection?
A. For legal reasons
B. To protect your license
C. To reduce the cost of repairs
D. To ensure safety

8. When does a rear-wheel braking skid take place?
A. Upon the locking of the rear wheels
B. When the rear and front wheels unlock
C. After the brake is in a different gear
D. When the rear wheels take enough weight

9. Which of the following is the right way to correct the braking skid of the wheels?
A. Accelerate to the left.
B. Drive to the right side of the lane.
C. Stay on the clutch.
D. Pull over.

10. Which of the following is not a step in vehicle inspection?
A. Go through the last inspection report.
B. Inspect the engine compartment.
C. Switch on the engine and inspect the vehicle components.
D. Turn on the engine before checking the lights.

11. Which of the following is true about engine belts?
A. Loose belts will lead to extreme heat.
B. Loose belts will not pump the water and/or turn the fan very well.
C. You can check the tightness of the V-belt by pressing it.
D. All of the above.

12. The total weight of a powered unit, which includes the trailer and the cargo, is called __________.
A. Gross Combination Weight (GCW)
B. Gross Vehicle Weight Rating (GVWR)
C. Gross Vehicle Weight (GVW)
D. Gross Combination Weight Rating (GCWR)

13. An axle weight is __________.
A. The weight of an ax
B. The weight that is transferred to the road or ground
C. The total weight of the load and the trailer
D. Th weight of the rear tires

14. Suspension systems are designed with ____________.
A. The user's weight capacity
B. Suspended devices for inspection
C. The manufacturer's weight rating
D. All of the above

15. What happens before the air pressure drops under 60 PSI?
A. A warning light will come on.
B. A buzzer will not come on.
C. Nothing happens.
D. None of the above.

16. How do you check if the service brakes are functional?
A. Drive forward at about 5 mph.
B. Quickly move the vehicle forward.
C. Do not wait for natural air pressure to build up.
D. Lightly handle the brakes.

17. When driving on a slippery or wet surface, you should never ___________.
A. Drive faster.
B. Stop driving as soon as possible.
C. Decrease to a crawling speed.
D. None of the above.

18. There are how many classes of fire?
A. Three
B. Four
C. Five
D. Six

19. How is fire from greasy liquids extinguished?
A. By cooling and quenching with water and dry chemicals
B. By cooling using carbon dioxide for heat shielding, using chemicals or smothering
C. By using nonconducting items like carbon dioxide or dry chemicals
D. By using a specialized extinguishing powder

20. Which of these statements is true about nitric acid?
A. It should be placed on top of other products.
B. It should not be placed on top of other products.
C. It can be packaged with less care.
D. The best way to keep it safe is to put it close to the driver's seat.

21. When is the best time to turn off the retarder?
A. If the cargo has great weight
B. On a slippery or wet road
C. If the vehicle is tilting to the right
D. If the vehicle is tilting to the left

22. Which of the following can be a form of distracted driving that leads to collisions?
A. Eating
B. Concentrating on the road
C. Glancing at the mirror
D. Stopping for a break

23. Which of the following will not help a driver avoid distraction?
A. Reviewing how electronics in the vehicle are operated before starting a trip
B. Programming radio stations before driving
C. Refusing to read or write while driving
D. Not adjusting mirrors in every part of the vehicle before the trip

24. If driving in fog is unavoidable, which of the following should you consider?
A. Reducing your speed before entering the fog
B. Switching off flashers
C. Watching only the front of the vehicle
D. Carefully driving past other vehicles

25. Which of the following is true about strong winds?
A. They make it easier to drive in a lane.
B. They make it difficult to keep your hands off the wheel.
C. Staying in your lane becomes difficult when driving out of tunnels.
D. Driving at 80 mph will protect you from a strong wind.

26. A driver who is drifting across the road, leaving the window open in cold weather and driving too fast or too slow is likely to be ___________.

A. On drugs

B. Sleepy

C. Ill

D. All of the above

27. What is the first thing you should do when stuck on railroad tracks?

A. Keep trying to drive your vehicle out.

B. Call the emergency line.

C. Get out of the vehicle and stay far from the tracks.

D. Shout for help.

28. How do you prevent your vehicle from rolling back?

A. Increase the gear.

B. Remove your foot from the brake after switching off the engine.

C. Use the service brake if necessary.

D. Apply the parking brake if necessary.

29. What happens if the ABS is not functioning?

A. A yellow light on the control panel will light up.

B. A green light will come on in front of the brake.

C. The brakes will stiffen.

D. The speed of the vehicle will increase.

30. Which of the following statements is false about backing up?

A. If possible, you should use a helper.

B. It is impossible to back up a large vehicle safely.

C. You should always get out and check if your path is safe.

D. You should use your mirrors.

31. Why should you not drive very close to the edge of the road?

A. You will get a poor view of the road.

B. Other vehicles will not be able to overtake you.

C. The top of your vehicle might hit a signpost or other objects.

D. None of the above.

32. Which of the following is defined as driving a vehicle with the desire to harm others on the road?
A. Drunk driving
B. Road rage
C. Anger driving
D. Aggressive driving

33. When operating a 60-foot truck at a speed of 50 mph, the most minimum following distance should be ___________.
A. 10 seconds
B. 5 seconds
C. 6 seconds
D. 7 seconds

34. Which of the following is true of air tank drains?
A. They are used to supply air pressure to the brakes.
B. They are meant to drain oil and water from the tank.
C. They are not necessary for school buses.
D. They should be used only in hot weather.

35. When there is an accident and there is an injured person you can help, you should ____________.
A. Do nothing until an ambulance arrives.
B. Give the injured person alcohol to ease the pain.
C. Help the injured person stand for at least three minutes.
D. Apply direct pressure on the injury to stop bleeding.

36. What should you do when inspecting the automatic transmission fluid?
A. Loosen all the nuts.
B. Fix leaks another time.
C. Patch up the door.
D. Allow the engine to run.

37. Which of the following should be done to prevent a rollover?
A. The cargo should be close to the front of the trailer.
B. The cargo should be near the back of the trailer.
C. The heaviest parts of the cargo should be positioned over the lightest parts.
D. The cargo should be near the ground.

38. After changing the tire, you should ___________.

A. Increase the tire pressure.

B. Decrease the tire pressure.

C. Wait for 45 minutes before driving.

D. Stop after driving for a while to be sure the nuts have not loosened.

39. Which of the following is true about seeing hazards?

A. It makes driving difficult.

B. They are never possible to see.

C. It helps you be prepared.

D. It poses a danger.

40. Which of the following is true regarding the height of the cargo's center of gravity?

A. It is highly important for the control of the vehicle.

B. It is unimportant unless you are driving in windy conditions.

C. It can only affect loaded vehicles.

D. None of the above.

41. To secure your cargo, you should check it in the first __ miles of the journey and after every __ miles.

A. 50; 100

B. 50, 150

C. 100, 150

D. 150, 200

42. Areas of the road that bleed tar __________.

A. Are often white

B. Become solid

C. Preserve tires

D. Are slippery

43. What step should you take if you are arrested for a non-parking violation?

A. Write an apology.

B. Never give up the keys to your vehicle.

C. Let your employer know.

D. None of the above.

44. How much space should you leave in front of your vehicle?
A. At least one second for every 10 feet of the length of your vehicle under 40 mph
B. At least five seconds for every 10 feet of the length of your vehicle under 40 mph
C. At least five seconds for every 15 feet of the length of your vehicle under 60 mph
D. At least five seconds for every 5 feet of the length of your vehicle under 30 mph

45. If you are driving a 40-foot vehicle, you should leave how many seconds of space between you and the vehicle in front of you?
A. Five
B. One
C. Four
D. Two

46. You are on a steep downgrade and driving properly in a low gear. Your safe speed is about 40 mph. What should you do if your speed gets to 40 mph?
A. Hold on to the brakes to limit your speed to 25 mph. Then let go of the brakes.
B. Keep your speed steady and hold on to the brakes when you get to 50 mph to slowly decrease your speed to 40 mph.
C. Hold on to the brakes to slowly reduce your speed to 35 mph. Then let go of the brakes.
D. Increase your speed to 60 mph.

47. Which of the following statements is correct about being vigilant while driving?
A. You should use prescribed drugs to keep you awake.
B. When you're tired, you should only stop for five minutes, then continue.
C. A cup of coffee is helpful for safe driving when you are feeling groggy.
D. None of the above.

48. If the surface of the road is packed with snow, you should decrease the speed by ________.
A. One-third
B. One-quarter
C. Half
D. None of the above

49. Which of the following statements is true about fire extinguishers?
A. They should always be charged.
B. You should not use a fire extinguisher until you have studied the instructions on it.
C. You should ascertain if the fire extinguisher is good for the type of fire that may likely arise.
D. All of the above.

50. Which of the following must be among your safety equipment?

A. A flashlight

B. Extra electrical fuses when your vehicle has circuit breakers

C. An extra mirror

D. A spare seat belt

18 Air Brakes Test

1. How many basic classifications of brakes are there?
A. Two
B. Three
C. Four
D. Five

2. Air brakes are sometimes called __________.
A. Three-in-one brakes
B. Automatic brakes
C. Manual brakes
D. Heavy brakes

3. Which of the following is used when there is a fault in the brake system?
A. The parking brake
B. The emergency brake
C. The service brake
D. The neutral brake

4. What happens when you try to stop a vehicle with the brake pedal?
A. The brake compressor kicks off as the engine gets started.
B. The air in the atmosphere is compressed.
C. The compressed air travels to the air reservoir through the compressor governor.
D. All of the above.

5. Which of the following is the center of the air brake system?
A. Storage tank
B. Air compressor
C. Brake pedal
D. Brake actuator

6. If you are driving on a highway and the air pressure is about 100 PSI, what will the air compressor governor do?
A. Stop the compressor from pumping air
B. Alert you to quickly drive down from the highway
C. Permit the compressor to continue pumping air
D. Turn on the emergency brake until the pressure reduces to 40 PSI

7. What is the normal air loss rate for single vehicles when testing for leaks?
A. Below 2 PSI per second
B. Below 2 PSI per minute
C. Above 3 PSI per minute
D. Above 5 PSI per minute

8. What will happen without the air reservoir?
A. The brake pedal will be cut into two.
B. The compressor will be limited in function.
C. The driver can still drive a long distance.
D. All of the above.

9. The brake actuator is ____________.
A. Connected directly to the tire
B. Also known as a piston triangle
C. The sole component that makes the brakes work
D. None of the above

10. Which of the following is a tiny component used for collecting separated specks of dirt in the air?
A. The dirt collector
B. The dirt separator
C. The air filter
D. The air drain

11. When you hit the brake pedals, which of the following supplies the pressure you need for operation?
A. Safety valve
B. Double valve
C. Triple valve
D. Supply valve

12. The air filter works on ________ before passing the air to the compressor.
A. Atmospheric air
B. Drainage pipe
C. Combustion engine
D. The pressure

13. What happens if the air pressure of your single vehicle goes below 3 PSI per minute?
A. There is an engine fire.
B. There is a gas leak.
C. Nothing happens.
D. There is too much air loss.

14. Which of the following is not a type of foundation brake?
A. Emergency brake
B. S-cam brake
C. Disc brake
D. Wedge brake

15. What happens in the foundation brakes when you release the brake pedals?
A. The S-cam rolls back and returns the brake shoes from the brake drum.
B. The brake shoes will go into the brake drum.
C. The shaft of the brake cam begins to twist.
D. The foundation brake gives way for the emergency brake.

16. The disc brake uses which of the following to function?
A. Water system
B. Air mechanism
C. Wind brake
D. The S-cam

17. Any time you observe that you need to apply more pressure to maintain the speed of your vehicle, it means ___________.
A. The brakes are working fine.
B. You need a bigger air storage tank.
C. The brakes of your vehicle are fading and may need replacement.
D. The brake pedal is very cold.

18. If you see a hazard ahead and need to warn other drivers behind you, which of these components should you use?
A. A warner
B. A stoplight switch
C. The mirror
D. A left turn

19. Which of the following statements is true about ABS?
A. It is a bonus to your normal brakes.
B. It is a replacement for your normal brakes.
C. It does not allow the wheels to roll up.
D. All of the above.

20. Dual air brake systems use how many sets of controls?
A. One
B. Two
C. Three
D. Four

21. What should you do to stop a vehicle that has a manual transmission?
A. Push the clutch before the engine RPM is idle.
B. Wait until the engine RPM is idle before pushing the clutch.
C. Do not worry about controlled pressure.
D. All of the above.

22. What do you do if the air pressure of your combination vehicle goes below 3 PSI per minute?
A. Keep driving for the next four to five trips.
B. Be aware that there is too much air leakage in your vehicle.
C. Refuel your vehicle.
D. Add more engine oil.

23. Which of the following is an advantage of air brakes?
A. They work better than other types of brakes.
B. The components of the air brakes are fixed where the chassis is designed.
C. The compressors of air brakes can also be used for horns and tire wipers.
D. All of the above.

24. Which of the following is true of air brakes?
A. They do not give you control of the vehicle.
B. They are not needed in heavy vehicles.
C. There is always wear and tear on components.
D. They have a flexible hose connection.

25. If any of the slack adjusters goes beyond one inch with the combined push rod, you should ________.

A. Adjust it.

B. Reduce it to half an inch.

C. Leave it.

D. None of the above.

19 Doubles and Triples Test

1. Which of the following should you expect when steering a double very fast?
A. To arrive at your destination very fast
B. A crack-the-whip tilting of your vehicle
C. The truck to gain more stability
D. All of the above

2. How much space should you allow between cars when driving doubles and triples?
A. A little
B. A good amount
C. Almost none
D. None of the above

3. What should you do before you change lanes?
A. Drive quickly.
B. Look only at the left.
C. Confirm that no vehicle is next to you.
D. All of the above.

4. What should you do when there is fog while you are driving?
A. Drive with greater skill.
B. Wait for it to clear.
C. Be very careful.
D. All of the above.

5. How should you check the air brakes of doubles?
A. Check the flow of air.
B. Open the closed valve of the emergency line.
C. Inspect the tractor control/protection valve.
D. All of the above.

6. If you want the air to get to the service line, you should ___________.
A. Steer the wheel.
B. Turn on the brake light.
C. Apply the hand brake.
D. Apply the parking brake.

7. Normally, the trailer supply control changes to emergency when the pressure range is ____________.

A. 15 to 20 PSI

B. 20 to 40 PSI

C. 45 to 50 PSI

D. 40 to 45 PSI

8. When you are inspecting the emergency brake, it should be _________.

A. Switched on

B. Switched off

C. Put in neutral

D. None of the above

9. What component should you use when performing a normal function in the inspection of the service brakes?

A. Hand valve

B. Foot pedal

C. Parking brake

D. Service brake

10. Which of the following is a tool for coupling one or two axles together?

A. Axle

B. Converter dolly

C. Twin trailer

D. Hand valve

11. If you are not sure of the brakes, what should you do while coupling?

A. Turn on the emergency brakes.

B. Turn on the service brakes.

C. Switch on the engine.

D. Chock the wheels.

12. Which of the following is not a step in picking up the converter dolly using the front semitrailer and the tractor?

A. Opening the pintle hook

B. Protecting the dolly in a lifted position

C. Bringing down the dolly support

D. Removing the dolly from the immediate trailer that is behind the tractor

13. How do you join the converter dolly with the back trailer?
A. Lock the trailer brakes.
B. Chock the wheels.
C. Set up the height of the trailer.
D. All of the above.

14. Why should the landing gear be raised when coupling?
A. To avoid damage when the trailer is driven forward
B. To increase the speed of the vehicle
C. To check how well the coupling has been done
D. All of the above

15. Which of the following is not a method of uncoupling the back trailer of doubles?
A. Setting the rig firmly on clear ground and within a strategic sequence
B. Pressing the parking brake
C. Connecting the air dolly to the electrical lines
D. Locking up the second trailer wheels if there are no spring brakes

16. When gently moving the tractor, which of the following will be pulled first?
A. The front trailer
B. The back trailer
C. The tractor
D. None of the above

17. You should press the __________ to stabilize the rig.
A. Parking brake
B. Emergency brake
C. Service brake
D. Landing gear

18. Which of the following is a process of uncoupling converter dollies?
A. Bringing down the dolly's landing gear
B. Removing the safety chains
C. Chocking the wheels
D. All of the above

19. If the dolly is beneath the back trailer, what should you do to the pintle hook?
A. Close it.
B. Open it.
C. Press it.
D. Uncouple it.

20. How do you couple the first semitrailer or tractor with the second or third trailer?
A. Use the same process for joining tractors to semitrailers.
B. Set the converter dolly in its place.
C. Ensure the safety of the second trailer at the back.
D. All of the above.

21. How do you inspect doubles and triples?
A. Inspect the lower or bottom fifth wheel.
B. Avoid greasing the system area.
C. Open up the jaws around the shank.
D. All of the above.

22. You should inspect the ________ fifth wheel.
A. Lower
B. Bottom
C. Upper
D. All of the above

23. To inspect the landing gear, you should ____________.
A. Totally lift the gear.
B. Check for and replace missing parts.
C. Search for air leaks.
D. All of the above.

24. The valves at the back of the last trailer should be _______ when you are inspecting doubles and triples.
A. Closed
B. Open
C. Up
D. Down

25. Ensure there is a/an__________ of the glad hands when connecting.
A. Open connection
B. Looseness
C. Secure connection
D. None of the above

20 Hazardous Materials (Hazmats) Test

1. Which of the following is an example of hazardous materials?
A. Explosives
B. Flammable liquids
C. Pesticides
D. All of the above

2. Which of the following is true of you as a transporter of hazardous materials?
A. You are allowed to drive every route.
B. You must not drink alcohol, even in your home.
C. You are restricted to driving certain routes.
D. You are a potential danger to every road user.

3. Which of the following is an example of a flammable liquid?
A. Water
B. Fuel
C. Arsenic
D. Asbestos

4. Buses are never allowed to carry more than how many pounds of Class 6 materials?
A. 50
B. 100
C. 150
D. 200

5. Battery acid and hydrochloric acid are which class of hazardous materials?
A. Class 1
B. Class 3
C. Class 5
D. Class 8

6. Hazmat regulations instruct shippers to __________.
A. Not carry radioactive elements.
B. Learn the necessity of hazmats.
C. Place a warning on every package the drivers are transporting.
D. All of the above.

7. What should you use in the identification of hazardous materials?
A. Markings
B. Naming
C. Scribbling
D. Painting

8. If the capacity of the portable tank is less than 1,000 gallons, the shipping name should not be below _____ inches tall.
A. 1
B. 2
C. 3
D. 4

9. What should you do when you are loading hazardous materials?
A. Be circumspect when handling containers that are made up of hazmats.
B. Avoid using materials that may puncture or damage the package.
C. Stay within 25 feet of the tank.
D. All of the above.

10. Which of the following statements is not true about the loading and unloading of hazardous materials?
A. You should keep materials far from heat sources because most hazardous materials are flammable.
B. You should seek to identify if there are leaks or broken containers.
C. You should fix leaks after the first shipment.
D. Leaks mean big trouble.

11. Which of the following may likely lead to an explosion if not well braced in their containers?
A. Gases and oxidizers
B. Explosives and flammable liquids
C. Corrosives and flammable solids
D. All of the above

12. Bulk packaging that is fixed to a truck is called a __________.
A. Vehicle tank
B. Cargo tank
C. Fixed tank
D. Trailer tank

13. Every portable tank should have __________.

A. The lessee's name

B. The driver's name

C. The name of the destination

D. A spare tank

14. The key stakeholders for the transportation of hazardous materials are ________.

A. The shipper, the carrier and the driver

B. The driver, the passenger and the police

C. The driver and the passenger

D. The driver and the students

15. The carrier is __________.

A. The driver of the vehicle

B. The party that transports products from the shipper to the final destination

C. The party that accepts improper shipment

D. Not in charge of accounting to the shipper and the appropriate government agencies when accidents or incidents occur

16. Which of the following tools is used to provide information on the type of hazardous materials involved when there is an accident?

A. The radio

B. The green box

C. The shipping paper

D. The placard

17. Which of the following is an example of a shipping paper?

A. Shipping orders

B. Bills of lading

C. Manifests

D. All of the above

18. Placards are placed on how many sides of the vehicle?

A. One

B. Two

C. Three

D. Four

19. Bulk packaging and cargo tanks place the identification number of packages on ___________.

A. Placards

B. Orange panels

C. White square-on-point

D. All of the above

20. Pesticide and arsenic are examples of which class of hazardous materials?

A. Class 3

B. Class 5

C. Class 6

D. Class 9

21. A freight container can be used how many times?

A. Twice

B. Three times

C. Multiple times

D. Once

22. How should Class 4 and 5 materials be kept?

A. In open areas

B. Secured

C. In a wet place

D. All of the above

23. Which of the following statements is not true about corrosive materials?

A. They should be kept far from heat.

B. They should be loaded one after the other alongside explosives.

C. They should be placed upright.

D. None of the above.

24. How far away should your vehicle be parked when there is an open fire, a bridge, a building, a tunnel or a multitude of people?

A. Not more than 250 feet

B. Less than 100 feet

C. Farther than 300 feet

D. Exactly 50 feet

25. A freight container can contain how many cubic feet in volume of gas or liquid?

A. 64

B. 664

C. 1,090

D. 98

21 Combination Vehicles Test

1. How do you prevent rollover risk?
A. Drive slowly in curves.
B. Stop abruptly when there is danger in front of your vehicle.
C. Increase your speed to attain balance.
D. Quickly change lanes when necessary.

2. You should leave at least one second for every 10 feet of the length of your vehicle at a speed under how many miles per hour?
A. 10
B. 20
C. 30
D. 40

3. To prevent skidding of a vehicle, you should ___________.
A. Look at the mirrors.
B. Avoid the brakes when getting back in traction.
C. Be very observant.
D. All of the above.

4. Longer trailers have _________ cheating compared to shorter ones.
A. Less
B. More
C. Better
D. No

5. When you want to make sure air goes to all parts of your vehicle, you should use __________.
A. The service line
B. The circular line
C. The projection line
D. The steering wheel

6. Why should you not engage the trailer hand valve when driving?
A. It may turn on the rear light.
B. It may later skid.
C. The vehicle will stop moving.
D. It decreases the speed of your vehicle

7. Which of the following maintains the air in the air brake system if there are leaks in the trailer?
A. The tractor protection control valve
B. The trailer hand valve
C. The air tank
D. The trailer air supply control

8. The tractor protection control valve causes air to circulate through the __________.
A. Service line
B. Parking line
C. Emergency line
D. All of the above

9. How many types of air lines are there?
A. Two
B. Three
C. Four
D. Five

10. Which of the following is true about the service air line?
A. It is also called the control line.
B. It is linked to the relay valve.
C. It transfers air.
D. All of the above.

11. The emergency air line is also known as the __________.
A. Control line
B. Supply line
C. Signal line
D. Urgency air line

12. What tool should you use to couple the service and emergency lines in the truck/tractor with the trailer?
A. The glad hands
B. The tractor protection valve
C. The shut-off valve
D. None of the above

13. When should dummy couplers be used?
A. Only in winter
B. When the emergency and service lines are not connected to the trailer
C. When the emergency line is connected to the trailer
D. When the service line is connected to the trailer

14. Which of the following is not applicable to the supervision of the fifth wheel?
A. You should look for destroyed or missing parts.
B. You should open the fifth wheel if it is sliding.
C. You should check to see if the fifth wheel is well greased to avoid frictional steering problems.
D. You should check to see if the kingpin is damaged or out of shape.

15. How should you set the position of the vehicle?
A. Set the trailer directly behind the tractor.
B. Put the trailer adjacent to the tractor.
C. Position it so you don't have to worry about the side mirrors.
D. All of the above.

16. When coupling combination vehicles, what will happen if the trailer is very low?
A. The load will fall over.
B. The vehicle will stall.
C. The tractor may hit the trailer.
D. None of the above.

17. Which of the following is not a procedure for providing air to trailers?
A. Pressing the air supply button
B. Switching the valve control from emergency to normal mode
C. Waiting for the air pressure to become normal
D. None of the above

18. When backing the tractor under a trailer, you should ____________.
A. Drive slowly.
B. Steer the wheel to the left.
C. Steer the wheel to the right.
D. Use the parking brake.

19. When lifting the landing gear so it becomes weightless, you should __________.
A. Increase the gear range.
B. Start from the lowest point again.
C. Stop lifting the gear.
D. Apply the parking brake.

20. Which of the following is not true about setting the rig in the right position?
A. You should know not all surfaces are good supports for the weight of trailers.
B. You should establish an alignment between the tractor and the trailer.
C. You should check the parking area and ensure it is strong enough to take the weight and size of the trailer.
D. You should use light surfaces when setting the rig.

21. When uncoupling, why should you chock the trailer wheels?
A. To decrease the speed of the vehicle
B. To make the wheels work better
C. To prevent movement
D. To increase air flow

22. Which of the following is not part of the process for disconnecting the electrical cable from the air lines?
A. Hanging the electrical cables
B. Using accepted lines
C. Setting apart the air lines from the trailer
D. Fixing the air line glad hands with the emergency line

23. The fifth wheel is uncoupled by _________.
A. Lifting the release handle
B. Putting the release handle on an open position
C. Placing your legs far from the back of the trailer wheels
D. All of the above

24. When pulling the tractor away from the trailer, what should you do?
A. Pull the tractor.
B. Not stop pulling the tractor until the fifth wheel is pulled from the base of the trailer.
C. Pause if the trailer is on the tractor's frame.
D. All of the above.

25. The tractor protection valve control is also known as the __________.

A. Trailer emergency valve

B. Security valve

C. Tractor supply valve

D. Trailer control valve

22 Tanker License Test

1. A tanker license is needed for portable tanks with a capacity of how many gallons or more?
A. 100
B. 119
C. 200
D. 1,000

2. Why do you need to inspect your tanker?
A. For the safety of the liquid contents
B. For the safety of the gaseous contents
C. For safe driving
D. All of the above

3. Which of the following should you consider when driving tanker vehicles?
A. Outage
B. Surge
C. Center of gravity
D. All of the above

4. Due to a greater center of gravity, the top of the vehicle is ____________.
A. Light
B. Heavy
C. Low
D. Curved

5. Which of the following is not true about a liquid surge?
A. It will affect your control.
B. It is the forward and backward movement of the liquid.
C. It does not occur from a wave.
D. Drivers need to be careful while driving liquids.

6. Liquid overflow is __________.
A. Rare
B. Mandatory
C. Dangerous
D. Very important

7. Excess weight should be ___________.
A. Spread to the back of your vehicle
B. Spread to the front of your vehicle
C. Avoided
D. All of the above

8. Which of the following is not true about baffled tanks?
A. They have holes.
B. They help in controlling the vehicle.
C. Their holes are for the passage of air.
D. All of the above.

9. When is it most appropriate to use unbaffled tankers?
A. When transporting all kinds of liquids
B. When transporting food products like milk
C. When transporting fuel
D. When transporting water

10. Expect a very ________ movement of the unbaffled tanker when transporting liquids.
A. Low
B. High
C. Balanced
D. Vertical

11. Which of the following is true about liquids?
A. They do not have the same expansion rate.
B. They do not expand.
C. They need to be filled to the brim of the tanker.
D. All of the above.

12. What should you do if a liquid is dense?
A. Fill up the tank.
B. Do not fill up the tank.
C. Keep the tank open.
D. Allow the liquid to cool before driving.

13. Which of the following does not determine the quantity of liquid you can load?
A. Weight of the liquid
B. Speed of the vehicle
C. Expansivity of the liquid
D. Legal limitations

14. You need to drive smoothly because of the tendency of ________ and ________.
A. Liquid surge; high center of gravity
B. Height; speed of your vehicle
C. Time; speed
D. Distance; time

15. It may take a long time to stop a truck with a/an __________.
A. Empty tank
B. Full tank
C. Half-full tank
D. All of the above

16. How should you drive on roads with many curves?
A. Carefully
B. Vigilantly
C. Slowly
D. All of the above

17. Skidding is a result of ________.
A. Too much acceleration
B. Overbraking
C. Overloading
D. All of the above

18. When inspecting for leaks, where should you check?
A. The outer part of the tank
B. The valves
C. The pipes
D. All of the above

19. When inspecting the valves, you should focus on the _________.
A. Intake
B. Discharge
C. Shut off
D. All of the above

20. The manhole covers should be __________.

A. Properly opened

B. Properly washed

C. Properly closed

D. None of the above

21. What should you do when a tank vehicle's valve and manhole covers are not closed?

A. Drive the loaded vehicle to the nearest workshop.

B. Transport all packages in just one trip.

C. Refrain from transporting packages.

D. All of the above.

22. When you find yourself in traffic, you should do which of the following?

A. Wait for enough of a gap so you can merge.

B. Rush into the next lane.

C. Call the police.

D. None of the above.

23. What should you do when driving your tanker in a curve?

A. Speed up.

B. Abruptly stop.

C. Push your vehicle hard through the curve.

D. Move the gear to the right downshift.

24. When should you park your vehicle?

A. Where backing up is easy

B. Where there are trucks

C. Only on asphalt

D. Only on gravel

25. Why should the ignition key be taken out before performing a pre-trip inspection on a tanker?

A. Someone might mistakenly drive the vehicle and crush whoever is under it.

B. It shows you are the vehicle owner.

C. You cannot do an inspection without the key.

D. None of the above.

23 Passenger Transport Test

1. Safety is ___________.
A. Not a necessity
B. Only required for tankers
C. A priority
D. Only needed when hazardous materials are in the vehicle

2. What should you do with the previous driver's inspection report?
A. Read through it before driving.
B. Keep it underneath your seat.
C. Sign it immediately.
D. All of the above.

3. Which of the following vehicle parts should you check?
A. The parking brake
B. The service brake
C. The horn
D. All of the above

4. The wiper _________.
A. Is a nonessential component and should not be inspected
B. Should be checked only when rain is about to fall
C. Is very important and should always be inspected
D. All of the above

5. The railing is where on a vehicle?
A. The interior
B. The exterior
C. Nowhere
D. None of the above

6. When should the floor covering be inspected?
A. After loading
B. After driving
C. Before loading
D. When going on a long-distance drive

7. Which of the following is true about roof hatches?
A. You should always leave the roof hatches open.
B. You can partially close up the roof hatches.
C. Roof hatches are meant to prevent air from coming in.
D. All of the above.

8. Tires and electrical fuses need ________.
A. Spare parts
B. Seat belts
C. Greasing
D. None of the above

9. If your seat belts are worn out, what should you do?
A. Ignore them for at least one more month.
B. Sew them up.
C. Replace them with new ones immediately.
D. Avoid ever using seat belts.

10. Which of the following is true about obstructions?
A. They may make passengers stumble.
B. They should be cleared out of the way.
C. They may lead to an accident.
D. All of the above.

11. The transportation of hazardous materials requires which of the following?
A. A hazmat license
B. A hazmat endorsement
C. Marking the hazmat materials
D. All of the above

12. Buses are allowed to transport which of the following?
A. Small arms ammunitions
B. ORM-D
C. Emergency supplies
D. All of the above

13. You are not allowed to carry _________.
A. Radioactive materials near passenger seats
B. Explosives near seats
C. Solid poisons that weigh more than 100 pounds
D. All of the above

14. The sum total weight of hazardous materials should not be more than ________.
A. 500 pounds
B. 200 pounds
C. 400 pounds
D. 100 pounds

15. The standee line is ___________.
A. Where the driver should stand and monitor passengers
B. The line that separates the driver from the passengers
C. The line that marks permissible hazardous materials
D. None of the above

16. Which of the following statements is true?
A. No passenger should stand behind the driver's seat.
B. Only teenagers are allowed to stand on the standee line.
C. Only adults are allowed to stand on the standee line.
D. All of the above.

17. Which of the following statements is not true?
A. It is wrong to announce the number of the vehicle when arriving at the final destination.
B. It is all right to stop without telling your passengers the reason.
C. You should not ever announce the destination.
D. All of the above.

18. Which of the following should you do before beginning a trip?
A. State all the vehicle rules.
B. Answer questions.
C. Make sure all passengers are seated.
D. All of the above.

19. To maintain your eyes on the road, use _______ to monitor what the passengers are doing.
A. The windshield
B. The service brake
C. The mirrors
D. All of the above

20. If you encounter a belligerent passenger, you should _______.

A. Shout at the passenger.

B. Drop the passenger off immediately.

C. Carefully handle the situation.

D. None of the above.

21. Which of the following statements is not true about handling bus crashes?

A. Traffic signs are enough to guide drivers.

B. Sometimes you may not be the driver at fault.

C. You should use discretion.

D. You need to prepare for the unexpected.

22. _________ and ________ are the greatest enemies to driving safety.

A. Roads, weather

B. Drivers, passengers

C. Slippery roads, speed

D. Traffic, roads

23. When stopping at railroad-highway crossings, you should do which of the following?

A. Apply the brake when your bus is about 15 to 50 feet from the crossing.

B. Not be too quick to jump into a lane.

C. Watch out for trains.

D. All of the above.

24. After a train passes, you should _________.

A. Quickly jump into the lane.

B. Wait to be sure a space is clear.

C. Turn on the emergency brake.

D. All of the above.

25. Which of the following is advised when you are in any stage of the transporting of passengers?

A. Do not fuel a bus if there are riders in the vehicle.

B. Do not be carried away with chitchat.

C. Never push or tow a faulty bus when the passengers are on board.

D. All of the above.

24 School Bus Endorsement Test

1. Which of the following is not the right way to manage students who are misbehaving?
A. Remind the offender(s) of the rules and regulations.
B. Switch conflicting parties from their seats.
C. Use your own disciplinary measures for the situation.
D. Stop the bus.

2. Students should always be ____________.
A. Reminded of the rules
B. Informed of the rules
C. Properly monitored using the mirror
D. All of the above

3. Which of the following is a possible step to take when there is an offender on the bus?
A. Call the school administration.
B. Call the police.
C. Have the student get off at the official bus stop.
D. All of the above.

4. How should you approach a designated bus stop?
A. With caution
B. With vigilance
C. Using the mirrors
D. All of the above

5. Which of the following is a step you should take when arriving at your destination?
A. Continue to check the mirrors
B. Stop the school bus 10 feet from the bus stop
C. Do not set the transmission in neutral or park
D. All of the above

6. When should the warning lights be turned on?
A. After arriving at the bus stop
B. Never
C. When approaching the bus stop
D. Only when loading

7. You should stop the school bus how many feet from the bus stop?

A. 3

B. 10

C. 25

D. 50

8. Where should you stay when parking?

A. Very close to the side of the driving lane

B. In the center of the road

C. Far from the driving lane

D. None of the above

9. Which of the following should you do when stopping at a school bus stop?

A. Wait for clearance.

B. Stay very close to the side of the lane.

C. Set the transmission in neutral.

D. All of the above.

10. When should students enter the bus?

A. When the driver signals

B. When they feel like it

C. When the bus is still on

D. As soon as they see the bus

11. When there's a missing student, you should __________.

A. Not drive away immediately

B. Check around and beneath the vehicle

C. Call the parents

D. All of the above

12. What should you do if the loading area is dark?

A. Switch off the dome light.

B. Turn on the dome light.

C. Load the students in the dark.

D. Turn on music.

13. Which of the following should you not do if the students will cross the road?
A. Walk about 45 feet.
B. Check if the red lights are flashing.
C. Look in all directions.
D. None of the above.

14. What should you do when you arrive at a school as your final destination?
A. Go inside to be sure that there are not any students who have not unloaded.
B. Ensure that no students are returning to the bus.
C. Count all the students and be sure none are missing.
D. All of the above

15. After accounting for each student, you should ____________.
A. Let go of the parking brake.
B. Switch on the alternating flashers.
C. Switch off the left turn signal.
D. Do not bother looking at the mirrors.

16. Which of the following is any area where schoolchildren are likely to receive the greatest injury from their bus or another bus?
A. Impact area
B. Danger zone
C. Injury line
D. All of the above

17. The first _______ feet of the danger zone are often more dangerous.
A. 30
B. 10
C. 20
D. 35

18. Which of the following is not true about mirrors?
A. They are optional components of your vehicle.
B. They help you see what is behind the bus.
C. They help you prevent collisions with other vehicles.
D. They are necessary for backing up.

19. A regular school bus's blind spot is __________.
A. 10 to 30 feet
B. 40 to 45 feet
C. 50 to 150 feet
D. 20 to 40 feet

20. Using the outside flat mirrors gives you how much of a view of the back of your bus?
A. The length of 7 buses
B. The length of 4 buses
C. The length of 10 buses
D. The length of 20 buses

21. Which of the following gives you a panoramic view of what is happening on the left and right sides of your environment outside the bus?
A. The outside flat mirrors
B. The outside convex mirrors
C. The outside crossover mirrors
D. The overhead rearview mirror

22. Which of the following gives you a view of the front part of the back tires where they make contact with the ground?
A. The outside flat mirrors
B. The outside convex mirrors
C. The outside crossover mirrors
D. The overhead rearview mirror

23. Properly adjusting the outside crossover mirrors will help you have a view of ___________.
A. Every zone ahead of the bus
B. All the front tires where they do not make contact with the ground
C. Areas from the bumper to where you have a direct line of sight
D. All of the above

24. When you use the overhead rearview mirror, where is there a blind spot?
A. As far as 400 feet behind the bus
B. At the back of the driver's seat
C. At the back of the bumper
D. All of the above

25. There are how many types of crossings?

A. Three

B. Five

C. Four

D. None of the above

17 Answers for General Knowledge Test

(1) (B) A danger.

Before you go under an overhead object, ensure you have overhead clearance. Do not rely on the heights written on bridges and overpasses. They may have been repaved and had the clearance reduced. Also, weight affects overhead clearance. A loaded cargo van is not as high as an empty one. If there are no warning signs on overhead objects, go slowly.

(2) (B) Very small when the vehicle is fully loaded.

You need to always pay attention to the space below your vehicle. A heavily loaded vehicle often has a smaller space below. This becomes problematic when the vehicle is driving on a dirt road or on uncemented/unpaved land. Be careful along these paths so that you don't stall. Do not take a chance on getting hung up in the middle of the road or in drainage channels.

(3) (C) Setting the back of your vehicle near the curb.

It is good to have space at the sides of trucks and buses. Large vehicles are susceptible to hitting objects or other vehicles when turning. This is often due to off-tracking and wide turning. When turning to the right, make it slow. If you are turning to the right, turn wide completely.

(4) (D) Could help avoid a crash.

An ABS can help you avoid a crash by giving you more control of the vehicle.

(5) (A) Total stopping distance.

Perception distance is the gap between the moment your eyes recognize the information of seeing the hazard to when the brain receives the warning. Reaction distance is the time gap between the command given by the brain and the actual placing of your foot on the pedal. Braking distance is the time it takes for the result of holding the brake pedal to actualize.

(6) (B) Control line.

The service line is also called the signal or control line. It transfers air around the air brake system. The service line pressure is based on the effort you exert on the brake or hand valve. The service line is linked to the relay valves, which facilitate the quicker application of trailer brakes.

(7) (D) To ensure safety.

Inspection is very important because it ensures the safety of you, your vehicle, your passengers and other road users.

(8) (A) Upon the locking of the rear wheels.

Rolling wheels do not have as much traction as locked wheels. This makes the back wheels slide backward when trying to meet up with the front wheels. A bus or straight vehicle will spin out in a sideways direction. With a towing vehicle, the skidding of the wheels will make the trailer move the towing vehicle to the side and jackknife.

(9) (C) Stay on the clutch.

The most common type of skid is when the back wheels lose traction because of too much acceleration or braking. Skids that result from acceleration often take place on ice or snow. Taking your foot off the accelerator can resolve the issue. However, if it is slippery, you should have the clutch pushed in.

(10) (D) Turn on the engine before checking the lights.

This is not a step in vehicle inspection. There are seven basic steps. The first is going through the last vehicle report. The second step is inspecting the engine compartment. The third step is switching on the engine and inspecting the vehicle components. The fourth step is turning off the engine before checking the lights. The fifth step is performing a walk-around inspection. The sixth step is checking the signal lights. The seventh step is turning on the engine and checking for hydraulic leaks and the status of the brake system.

(11) (D) All of the above.

When driving in hot weather, inspect the tightness of your vehicle's V-belt by pressing it. When the belts are loose, it affects the performance of the engine because the water pump and the cooling fan will not work very well. This will lead to overheating of the engine and the vehicle. It is also good to check the belts for cracks and wear.

(12) (A) Gross Combination Weight (GCW).

A Gross Combination Weight (GCW) is the sum of the weight of a powered unit, which includes the trailer and the cargo. Gross Combination Weight Rating (GCWR) is the highest GCW that the manufacturer specifies for a particular combination of vehicles, including the load. The Gross Vehicle Weight (GVW) is the sum of the weight of just one vehicle, including the load. The Gross Vehicle Weight Rating or GVWR is the highest GVW that the manufacturer specifies for a single vehicle that includes the load.

(13) (B) The weight that is transferred to the road or the ground.

An axle weight is the weight that a single axle or set of axles sends to the ground. It is your responsibility to ensure there is balance and no overload in any component of the vehicle. Keep the weight well distributed and reduce the load if needed.

(14) (C) The manufacturer's weight rating.

Suspension systems are designed with the manufacturer's weight capacity. It is your responsibility to identify this rating. Drive and maintain your vehicle according to this capacity rating. It is specific to every vehicle and should not be assumed to be the same for every vehicle.

(15) (A) A warning light will come on.

A warning light and buzzer turn on before the air pressure falls under 60 PSI in the air systems. When this occurs while you are driving, stop immediately and park safely. If one of the air systems does not have enough pressure, the front or back brakes will be less functional. In other words, it will take you longer to stop your vehicle.

(16) (A) Drive forward at about 5 mph.

To check service brakes, wait for normal air pressure to build up, then let go of the parking brake. Slowly drive the vehicle forward at about 5 mph. Then firmly hold on to the brakes with the brake pedal. Notice any pulling or tilting toward an angle or delay in the final stop.

(17) (A) Drive faster.

It is not easy to stop on slippery surfaces, and it is very difficult to turn without skidding if the surface of the road is slick. The stopping distance of wet roads is sometimes about double that of dry roads. This is why you should never drive faster in such situations. Instead, decrease speed by approximately one-third of the general stopping distance. Therefore, if the stopping distance is 75 mph, you should slow down to 50 mph.

(18) (B) Four.

Fire is classified into four classes—A, B, C and D. Class A includes wood, paper and ordinary combustions. Class B includes greasy liquids, such as gasoline, oil and grease. Class C includes electrical equipment. Class D includes combustible metals. Each of these classes has its own way of being extinguished.

(19) (B) By cooling using carbon dioxide for heat shielding, using chemicals or smothering.

A Class A fire is extinguished by cooling and quenching with water and chemicals. A Class B fire (greasy liquids) is extinguished by cooling using carbon dioxide for heat shielding, using chemicals or smothering. A Class C fire is extinguished by using nonconducting items, like carbon dioxide or dry chemicals. A Class D fire is extinguished by using a specialized extinguishing powder.

(20) (B) It should not be placed on top of other products.

This statement is true. Nitric acid is a hazardous material. It should not be placed on top of other materials.

(21) (B) On a slippery or wet road.

Retarders are built into some vehicles and are best to use on a slippery or wet road. They help reduce speed and make it unnecessary to use brakes.

(22) (A) Eating.

Different activities can distract a driver. These activities include adjusting the radio; drinking, smoking or eating; chatting with the passengers on board; concentrating on billboards and people; making a phone call; engaging in mental preoccupations like daydreaming and worrying; and picking up fallen items. When you are not focused when driving, the lives of everyone are endangered.

(23) (D) Not adjusting mirrors in every part of the vehicle before the trip.
Not adjusting mirrors in every part of the vehicle before the trip is dangerous. Mirrors should be adjusted before the trip. This will ensure that you are not distracted while driving. Also, do not read, write, smoke or drink while driving. Avoid intense discussions. Taking your hands off the steering wheel can be dangerous. Therefore, radio stations should also be programmed before the trip.

(24) (A) Reducing your speed before entering the fog.
It is better to wait for the fog to clear, but if you must drive through it, reduce your speed, avoid overtaking other vehicles, turn on the low-beam lights for visibility, watch out for vehicles at the side of the road, adhere to warning signs and turn on the four-way flashers. Since there is not much visibility, use your ears to check for traffic.

(25) (C) Staying in your lane becomes difficult when driving out of tunnels.
If you are caught in strong winds, staying in your lane may become difficult. You should hold tight to the steering wheel to retain maximum control of it. Slow down and do not fight the wind. You may have to park somewhere safe and wait for the wind to die down. Reach out to your dispatcher to find out what to do next.

(26) (D) All of the above.
When driving, you have to beware of impaired drivers who are drowsy, have taken drugs, are drunk or are sick. They may be hazards and should be avoided. You will know them by the way they drive. They often drift or weave across the road, stop at an inappropriate time or are too fast or too slow.

(27) (C) Get out of the vehicle and stay far from the tracks.
A railroad track is a very dangerous place to get stuck in. The unpredictable can occur within the blink of an eye. Therefore, you should get out of the vehicle immediately before you attempt to do anything else. Move far from the tracks. Look around for a sign or signal to know if there is an emergency notification line. Then call 911 or any other emergency line.

(28) (D) Apply the parking brake if necessary.
You should apply the parking brake if necessary to prevent the vehicle from rolling back.

(29) (A) A yellow light on the control panel will light up.
The antilock braking system (ABS) is an add-on to the brakes. It does not increase or reduce the speed. The brake will still function effectively if there is no ABS. For new systems, the yellow light indicator turns on at the start-up to check the bulb, then goes off immediately. For older systems, the light will remain on until you have driven as far as five miles.

(30) (B) It is impossible to back up a large vehicle safely.
This statement is false about backing up. Backing up is very dangerous because you cannot see everything at the rear. Avoid backing up as much as possible. If it is necessary to back up, stay in the right position, use your mirrors, check if your path is safe, back up slowly, use a helper and back toward the angle of the driver.

(31) (C) The top of your vehicle might hit a signpost or other objects.
When you drive too close to the edge of the road, your vehicle may shift toward the roadside. This can make the top of your vehicle hit a signpost, tree limb or other objects. It becomes difficult to drive when crossing the shoulder and leaving the road. Therefore it's advisable to drive in the center of your lane.

(32) (B) Road rage.
Road rage is the driving of a vehicle with the desire to harm others on the road. Road rage is not a mistake. It is intentional and malicious. It is unlike driving when drunk, which may be circumstantial.

(33) (D) 7 seconds.
The rule of keeping a safe distance specifies at least one second for every vehicle length at a speed under 40 mph. When the speed goes up, increase the seconds by one second to enable safety. Therefore, if the speed is over 40 mph, then you should add one second to make up for the additional speed. Sixty feet makes six seconds plus one second for the extra 10 mph.

(34) (B) They are meant to drain oil and water from the tank.
This statement is true. Air tanks have a draining valve at the end to remove water and compressor oil residuals that may be in the tank. Failure to drain the water and oil may lead to faults in the air brake system if the liquids freeze during cold weather. Draining of the air tank can be done manually or automatically.

(35) (D) Apply direct pressure on the injury to stop bleeding.
You can relieve pain and the extent of the injury by providing warmth for the injured person, exerting direct pressure against the wound to stop bleeding and keeping the person still if it is a serious injury. You can move a seriously injured person only if there's an imminent danger of fire.

(36) (D) Allow the engine to run.
The automatic transmission fluid level is checked when inspecting the engine compartment. Allowing the engine to continue running will help you ascertain the fluid level and other aspects that may not be identifiable if the engine is off. Other things to check in the engine compartment are leaks, cracked wirings, the tightness of the belt and the extent of wear and tear.

(37) (D) The cargo should be near the ground.
Rollovers are a result of turning very fast. You should slow down when driving on or off ramps and in curves. Do not be too fast when changing lanes, especially when there is a full load. Keeping the cargo low will keep the load centered on the rig. The load should be in the middle and not leaning toward the left or the right.

(38) (D) Stop after driving for a while to be sure the nuts have not loosened.

Checking the tightness of the nuts will help prevent unfortunate situations on the road. Drivers need to always take note of this before, during and after a trip. Therefore, it is necessary to stop after a while and check. When there is rust around the wheel, it is likely that the nuts are loose.

(39) (C) It helps you be prepared.
Seeing hazards gives you ample time to prepare and act before they cause problems that escalate into emergencies. Preparation prevents danger. Abruptly braking because of unforeseen hazards can lead to a crash. When you see the hazard ahead, you can slow down, use your mirrors and signal for a lane change.

(40) (A) It is highly important for the control of the vehicle.
If your vehicle's center of gravity is very high, it means that your vehicle can easily roll over. A vehicle with an unbalanced center of gravity will be in great danger in curves or situations that require swerving away from hazards. Therefore, you should spread the load out to reduce the center of gravity. The heaviest parts of the cargo should be below the lightest parts.

(41) (B) 50, 150.
Inspecting your cargo within 50 miles will help you ensure the safety and security of your equipment. Then you should check again after every 150 miles. It is not ideal to have your cargo unchecked until the end of your trip. A part may need adjustments along the way, and it is your duty to ensure repair before it escalates to a critical danger.

(42) (D) Are slippery.
When driving, beware of bleeding tar that often occurs when the surface of the road becomes too hot. These surfaces are very slippery. They are also dangerous for drivers because they may result in an accident if the driver is not skillful or vigilant or drives too fast.

(43) (C) Let your employer know.
You should avoid traffic violations as much as possible. However, if you break any of the traffic rules and regulations and are convicted, the best thing to do is tell your employer about it.

(44) (A) At least one second for every 10 feet of the length of your vehicle under 40 mph.
This rule of keeping a safe distance specifies at least one second for every 10 feet of the vehicle's length at a speed under 40 mph.

(45) (C) Four.
The rule of keeping a safe distance specifies at least one second for every 10 feet of the vehicle's length at a speed under 40 mph. When the speed goes up, increase the seconds by one second for safety. For example, if you are driving a 20-foot vehicle, leave a distance of two seconds.

(46) (C) Hold on to the brakes to slowly reduce your speed to 35 mph. Then let go of the brakes.
Using the brakes on a steep or long downgrade helps achieve a braking effect. If you have placed the vehicle in low gear, hold the brakes hard. If your speed goes down to about 5 mph below your

safe speed, then let go of the brakes. Hold the brakes for approximately three seconds. Keep repeating this until you reach the end of the downgrade.

(47) (D) None of the above.
When you feel sleepy when driving, the best thing to do is to stop and take a nap. A 20-minute nap is better than 20 cups of coffee.

(48) (C) Half.
If the road is covered with snow, decrease the driving speed to half. If the road is icy, slow to a crawling speed and stop driving if possible. The stopping distance on wet roads is sometimes about double that of dry roads. Decrease the speed by approximately one-third of the general stopping distance.

(49) (D) All of the above.
To prevent an escalation of fire, you must keep your fire extinguisher charged, study the instructions before using the extinguisher and determine what class of fire it can be used on. There are different classes of fire—Classes A, B, C and D. The fire may be a combination of two or more classes.

(50) (B) Extra electrical fuses when your vehicle has circuit breakers.
Before you embark on a trip, be sure that the necessary safety equipment is available. Some of these are three reflective triangles for emergency purposes, a well-charged fire extinguisher, an accident report kit, tools for changing tires, extra electrical fuses, etc. These items help reduce the danger of breaking down while on a trip, especially if there is no assistance or mechanic close by.

18 Answers for Air Brakes Test

(1) (A) Two.

There are two basic types of brakes—the disc brake and the drum brake. These brakes can be further grouped into vacuum brakes, air brakes, hydraulic brakes, etc. In disc brakes, the brake chamber winds the power screw by exerting pressure on the slack adjuster.

(2) (A) Three-in-one brakes.

Some people prefer to call the air brakes "three-in-one brakes" because they comprise three connected braking systems. These parts are the parking brake, which is used when attempting to pull over or parking; the service brake, which is what you use in your usual driving routine; and the emergency brake, which is used when there is a fault in the brake system.

(3) (B) The emergency brake.

The emergency brake uses both the service brake and parking brake to stop the vehicle when there is a brake system failure. It is mandatory for every bus, truck and other heavy vehicles to have emergency brakes.

(4) (D) All of the above.

Anytime you try to use the brake pedal to stop a vehicle, the brake compressor kicks off immediately after the engine is turned on. Then, there is compression of the air, which the compressor governor passes on to the reservoir. This goes on in cycles as air is accumulated in the reservoir.

(5) (B) Air compressor.

The air compressor is the main component of an air brake system that uses a belt drive. Its purpose is to compress air in the environment to a particular pressure and take it to the air reservoir or storage tank.

(6) (C) Permit the compressor to continue pumping air.

The air compressor governor ensures there is proper timing of air supply to the reservoir. For instance, when the atmospheric air has reached a maximum level of 125 PSI, the air governor makes sure there is continual pumping of air from the compressor. When the air pressure is about 100 PSI, the air compressor governor will permit the compressor to continue pumping air.

(7) (B) Below 2 PSI per minute.

When testing for air leakage rate, switch off the engine, let go of the parking brake and monitor the amount of time it takes for the air pressure to fall. Often, the air loss rate is below 2 PSI per minute if it is a single vehicle and below 3 PSI per minute if it is a combination vehicle. Proceed to exert 60 PSI and above using the brake pedal.

(8) (B) The compressor will be limited in function.

The compressor and the storage tank, or reservoir, are complementary in function. The compressor pumps in the extra air, while the storage tank keeps it until the stored air is in demand. It serves as a reservoir for atmospheric air, which has been compressed under high pressure. It is vital because it is what sustains the movement of a vehicle on a trip. If there is no air reservoir, the compressor will be limited in function.

(9) (D) None of the above.
The brake pedal is connected directly to the brake pedal. This is also known as the piston cylinder. Whenever you push the brake pedal, the brake actuator is what releases pressure on the brakes. Brake pedals need a brake actuator. It is the same way the compressor governor needs the storage tank or reservoir. They are all complementary in function.

(10) (A) The dirt collector.
The dirt collector is a tiny component that collects separated specks of dirt from the air filters. It precedes the triple valve. Dirt in the tank can make the vehicle develop unusual faults. The dirt collector ensures specks of dirt do not block the passage of air. Your air brake system needs clean air for efficiency.

(11) (C) Triple valve.
You cannot operate an air brake system without the triple valve; it is in charge of the air brake operations. It functions as a press and release mechanism for the brake. As you hit the brake pedals, the triple valve supplies the needed pressure, then it releases the pressure when you relax on the brake pedal.

(12) (A) Atmospheric air.
The air filter and dryer work on the atmospheric air before it is passed on to the air compressor. The air filters separate dust particles from the atmospheric air.

(13) (D) There is too much air loss.
After the pressure drops, the normal air loss rate for single vehicles is below 2 PSI per minute. However, if the air pressure of your single vehicle goes below 3 PSI per minute, it means there is too much air loss.

(14) (A) Emergency brake.
The wheels of your truck are operated with foundation brakes. The most common type is the S-cam brake. The other two types of foundation brakes are wedge brakes and disc brakes. They are all needed for the turning of the wheel. An emergency brake is not a type of foundation brake.

(15) (A) The S-cam rolls back and returns the brake shoes from the brake drum.
The S-cam operates when you apply the brake pedal. Applying the brake pedal allows air into the brake chamber, and the pressure from the air makes the rod protrude. This affects the slack adjuster, which twists the shaft of the brake cam and rolls the S-cam. When you release the brake pedals, the S-cam rolls back and returns the brake shoes from the brake drum.

(16) (B) Air mechanism.

Disc brakes use an air mechanism in the application of air pressure against a brake chamber in a power screw slack adjuster, which resembles the S-cam. The pressure from the impact of the brake chamber against the slack adjuster moves the power screw. Then, the power screw hits the rotor, which is found against a caliper brake lining pad.

(17) (C) The brakes of your vehicle are fading and may need replacement.

The more pressure you apply to keep up with speed, the more this is a sign that your brakes may be fading and need immediate replacement. A need for increased pressure may be due to mechanical faults, air leaks or brake misalignment. Before you replace the application pressure gauge, hold on to your lower gear to slow down.

(18) (B) A stoplight switch.

When you turn on the switch brakes, the stoplight switch puts on the brake lights. It serves as a warning to drivers behind you.

(19) (A) It is a bonus to your normal brakes.

An ABS helps prevent your wheels from locking themselves. When your ABS is malfunctioning, your normal brakes will still be available for use. However, you will have to quickly repair the ABS. Also, ABS does not necessarily shorten your stopping distance, but it does help you keep the vehicle under control during hard braking.

(20) (A) One.

Although the air brake system is dual, it uses just one set of controls. Each comes with separate air brake parts, such as hoses, brake lines, air tanks, etc. One takes over the front axle, while the other operates the back axle. Both provide air to the vehicle.

(21) (D) All of the above.

When you intend to stop your heavy vehicle normally, pull down the brake pedals. Use controlled pressure to arrive at a smooth stop. If the transmission is manual, wait until the engine RPM is idle before you push the clutch. Choose the starting gear after stopping.

(22) (B) Be aware that there is too much air leakage in your vehicle.

After the pressure drops, the normal air loss rate for single vehicles is below 2 PSI per minute. However, if the air pressure of your single vehicle goes below 3 PSI per minute, it means there is too much air loss.

(23) (D) All of the above.

Air brakes work better than other types of brakes. The components of the air brakes are fixed where the chassis is designed. The compressors of air brakes can also be used for car horns.

(24) (D) They have a flexible hose connection.

The hose connection of air brakes is flexible. When driving a vehicle with air brakes, the truck or bus can easily be controlled. This limits necessary stopping distance. Wear and tear of the system components is reduced.

(25) (A) Adjust it.

During an inspection, check the slack adjusters. Stop your vehicle on a level surface and make sure the wheel is not moving. You can chock it. During your walk-around inspection, check manual slack adjusters on S-cam brakes. Put on your gloves and pull the slack adjusters. If any of the slack adjusters goes beyond one inch with the combined push rod, then you need to adjust it.

19 Answers for Doubles and Triples Test

(1) (B) A crack-the-whip tilting of your vehicle.

If you steer your large vehicle too fast, you will experience a crack-the-whip tilting. Steering gently helps prevent this. Be gentle when pulling over and never make a sudden move. The crack-the-whip effect makes doubles and triples more unstable and likely to topple over than combination vehicles.

(2) (B) A good amount.

Doubles and triples occupy more space on the road than other vehicles do. They are longer and cannot be stopped abruptly. Therefore, you should allow a good amount of space between your vehicle and others.

(3) (C) Confirm that no vehicle is next to you.

It takes a few seconds before doubles and triples can be fully stopped. Therefore, do not follow other vehicles too closely. Before you change lanes, confirm that no vehicle is next to you.

(4) (D) All of the above.

Fog often leads to accidents. A double or triple vehicle requires greater driving skill in these adverse conditions. If conditions are not ideal for driving, you should either wait for the weather to clear up or be extra careful on the road. It is better to arrive at your destination late than to endanger yourself or others.

(5) (D) All of the above.

When inspecting the air brakes of doubles and triples, check the flow of air, inspect the tractor control/protection valve, the trailer's service brakes and emergency brakes. These steps are to be combined with the steps for inspecting combination vehicles.

(6) (C) Apply the hand brake.

If you want the air to get to the service line, apply the trailer's hand brake. To ensure there is a maximum flow of air around the trailers, engage the trailer's parking brake and/or stop the moving vehicle through chocking the wheel. After waiting for the air pressure to become normal, press the red switch, which supplies air to the emergency line.

(7) (B) 20 to 40 PSI.

The tractor protection valve control, which is also known as the trailer supply control, pops up and changes its position to emergency normal when air pressure is between 20 to 40 PSI or according to the manufacturer's manual.

(8) (A) Switched on.

When you are inspecting the trailer's emergency brakes, switch on the air brake system and ensure there is no obstruction to the rolling of the trailer. Then stop and bring out the air supply control. Or you can simply set it in an emergency position. Slowly pull out the trailer and the tractor to ensure that the trailer emergency is switched on.

(9) (B) Foot pedal.

When inspecting the trailer's service brakes, use the hand valve to inspect the trailer's brakes, but use your foot pedal if it is for a basic function. The foot pedal will mount air upon the service brakes and every wheel of the system.

(10) (B) Converter dolly.

A converter dolly is a tool for coupling one or two axles together, including a fifth wheel. You can couple the semitrailer to the back of a tractor-trailer. The second trailer should be behind the converter dolly. The second trailer should be behind the converter dolly.

(11) (D) Chock the wheels.

Coupling needs to be done without disturbances or distractions. Chocking the wheels will keep it from moving when coupling. So, if you doubt the functionality of the brakes, just chock the wheels to be on the safe side.

(12) (A) Opening the pintle hook.

This statement is false. When using the front semitrailer and the tractor to pick up the converter dolly, you should set the combination very close to the converter dolly, lock the pintle hook, protect the dolly in a lifted position, bring the nose of the rear semitrailer closer to the dolly, bring down the dolly support and remove the dolly from the immediate trailer that is behind the tractor.

(13) (D) All of the above.

When coupling the converter dolly with the back trailer, ensure that you have locked the trailer brakes and chocked the wheels. Set up the height of the trailer. The converter dolly should be set beneath the back trailer. Safely set the landing gear in a lifted position. Pull on the pin of the rear semitrailer. Ensure there is no gap in the lower and upper fifth wheels.

(14) (A) To avoid damage when the trailer is driven forward.

When coupling the converter dolly with the back trailer, the converter dolly should be set beneath the back trailer. Safely set the landing gear in a lifted position and above the ground to avoid damage when the trailer is driven forward. You do not want to incur damage to any part of your vehicle when coupling.

(15) (C) Connecting the air dolly to the electrical lines.

This statement is false. When uncoupling the back trailer of doubles, set the rig firmly on clear ground and within a strategic sequence. Press the parking brake. If your vehicle does not have spring brakes, lock up the second trailer's wheels. Release the rear semitrailer's landing gear. Lock up the shut-off behind the front trailer. Remove all connections between the air dolly and the electrical lines.

(16) (A) The front trailer.

When you gently move the tractor, the front trailer will be pulled first and then the dolly out of the back of the semitrailer. This is because the front trailer precedes the back trailer and will be the first to follow when being pulled.

(17) (A) Parking brake.

The process of uncoupling the back trailer involves setting the rig on clear ground and applying the parking brake to stabilize it. Also, if your vehicle does not have spring brakes, lock up the second trailer's wheels. Release the rear trailer's landing gear to remove the weight or pressure from the dolly.

(18) (D) All of the above.

When uncoupling a converter dolly, bring down the dolly's landing gear. Remove the safety chains. Press the converter's spring brakes or chock the wheels to prevent movement of the vehicle. Unlatch the pintle hook, which is on the front semitrailer. Gently drive away from the dolly.

(19) (A) Close it.

After chocking the wheels, you should unlatch the pintle hook, which is on the front semitrailer. If the dolly is beneath the back trailer, do not open the pintle hook. Expect the dolly bar to go up if you do that. This may result in injury.

(20) (D) All of the above.

When coupling the first semitrailer or tractor with the second or third trailer, apply the process for joining the tractor to semitrailers. Ensure the safety of the second trailer at the back, integrate the converter dolly with the front trailer and integrate the converter dolly with the back trailer.

(21) (A) Inspect the lower or bottom fifth wheel.

When inspecting doubles and triples, inspect the lower or bottom fifth wheel.

(22) (D) All of the above.

When inspecting doubles and triples, inspect the lower or bottom fifth wheel. Ensure that it is safely placed on the frame. Check for missing, leaking or faulty components. Sufficiently grease the system area. Identify and close up gaps that exist between the top and bottom fifth wheels. Inspect the upper or top fifth wheel.

(23) (D) All of the above.

When inspecting the landing gear, keep the gear totally lifted up. Check for any missing or damaged parts that need replacement or repair. Remember to securely maintain the crank handle. Watch out for air leaks. Quickly fix any leaking components.

(24) (A) Closed.

When inspecting doubles and triples, keep the valves closed. However, those at the back of the front trailer should be open, while the valves at the back of the last trailer should be closed. The valve stem of the converter dolly's air tank should be closed.

(25) (C) Secure connection.

When inspecting the components of doubles and triples, ensure all air lines are well fixed and the glad hands are securely connected.

20 Answers for Hazardous Materials (Hazmat)

(1) (D) All of the above.

Examples of hazardous materials are explosives, solids, gases, flammable liquids, poisons, radioactive substances and other harmful materials. Hazmats pose a critical risk to drivers, passengers and citizens. Therefore, the government at all levels regulates the acquisition and use of these hazardous materials.

(2) (C) You are restricted to driving certain routes.

For safety reasons, hazmat drivers are sometimes asked to drive certain routes. This is to protect pedestrians and other drivers from the risk of hazardous materials. A driver may be prohibited from driving roads that are slippery, curvy or potholed. Drivers of hazardous materials are prohibited from using certain substances, such as alcohol, while driving.

(3) (B) Fuel.

Acetone and fuel are examples of flammable liquids. They fall under Class 3 of hazardous materials.

(4) (B) 100.

Buses are never allowed to carry more than 100 pounds of Class 6 materials.

(5) (D) Class 8.

Class 8 hazardous materials are acids.

(6) (C) Place a warning on every package the drivers are transporting.

One major reason for hazmat regulations is to inform drivers of the risks. Placing a warning on every package drivers are transporting will help people identify the nature of hazardous material in the package.

(7) (A) Markings.

Markings help in identification of hazmats. They should be placed on portable tanks and cargo tanks and should highlight the shipping name of all contents on both sides.

(8) (A) 1.

The size of the shipping name should not be less than one inch tall if the container's capacity is less than 1,000 gallons.

(9) (D) All of the above.

Loading a hazardous product requires great care to prevent unforeseen incidents. While loading, be circumspect when handling containers that are made up of hazmats. Avoid using materials that may puncture or damage the package. Stay within 25 feet of the tank.

(10) (C) You should fix leaks after the first shipment.

This is not true. You should always fix leaks before commencing shipment. Keep materials far from heat sources because most hazardous materials are flammable. Heat is a potential danger to all people and assets nearby. Seek to identify if there are leaks or broken containers.

(11) (D) All of the above.

The following items should be packed tightly to avoid movement or sliding of the products when being transported: gases, oxidizers, flammable liquids and solids, explosives, corrosives, radioactive materials and poisons. Safely brace containers that house these products. Two or more substances may hit each other or fall and break open when the vehicle hits a bump or goes around a corner.

(12) (B) Cargo tank.

Bulk packaging that is fixed to a truck or trailer is called a cargo tank. These are permanently on the vehicle when you are loading or unloading.

(13) (A) The lessee's name.

Every portable tank should have the owner or lessee's name clearly labeled.

(14) (A) The shipper, the carrier and the driver.

There are three key stakeholders for the transportation of hazardous materials—the shipper, the carrier and the driver. The shipper is in charge of obtaining shipping papers and transporting goods. The carrier is in charge of proper documentation and accountability to agencies. The driver is mainly in charge of driving packages from one point to another.

(15) (B) The party that transports products from the shipper to the final destination.

The roles of the carrier include transporting products from the shipper to the final destination; ensuring proper documentation, marking, labeling and availability of necessary materials; and accounting to the shipper and the appropriate government agencies when accidents or incidents occur.

(16) (C) The shipping paper.

When there is an accident, you may be unable to talk and therefore find it difficult to communicate the types of hazardous materials that are in your vehicle. Shipping papers can alert emergency responders to the dangerous contents of your cargo.

(17) (D) All of the above.

Manifests, shipping orders and bills of lading are examples of shipping papers.

(18) (D) Four.

A vehicle that has been placarded should have no less than four identical placards on the front, back and sides of the truck.

(19) (D) All of the above.

Placards are used to signal other drivers of the presence of hazardous materials. They are signs situated outside a truck or tanker and on top of bulk packages that contain hazardous materials. Orange panels and white square-on-point can also show the identification number.

(20) (C) Class 6.

Examples of Class 6 materials are pesticides, arsenic, etc. Buses are never allowed to carry more than 100 pounds of Class 6 materials. Some poisonous substances are flammable and dangerous when inhaled. Therefore, warn others of the hazard.

(21) (C) Multiple times.

A freight container can be used multiple times. It is fabricated to allow the intact raising of its contents when loading or unloading for transport and is used to securely transport hazardous materials from one point to another.

(22) (B) Secured.

Class 4 hazardous materials are flammable gases. Class 5 hazardous materials are oxidizers and organic peroxide. Keep Class 4 and 5 hazardous materials in secured and enclosed containers because they are harmful or reactive when wet. Therefore, they should be very dry. They should be kept far from heat.

(23) (D) None of the above.

Corrosive materials are hazardous. When loading corrosive liquids, do not load them close to explosives, flammable solids, oxidizers and poisonous gases. This is to prevent potential dangers, such as explosions and fire.

(24) (C) Farther than 300 feet.

You should park farther than 300 feet from an open fire, a bridge, a building, a tunnel or a multitude of people.

(25) (A) 64.

A freight container can contain 64 cubic feet in volume of gas or liquid.

21 Answers for Combination Vehicles

(1) (A) Drive slowly in curves.

To keep your vehicle stable, drive slowly around curves and ensure the cargo is not far from the ground. Do not change lanes abruptly. Fast driving causes rollovers. Be gentle when pulling over and never make a sudden move.

(2) (D) 40.

The rule of keeping a safe distance specifies at least one second for every 10 feet of a vehicle's length at a speed under 40 mph. When the speed goes up, increase the seconds by one second for safety. If you are driving a 20-foot vehicle, the distance of time is a two-second gap. However, if the speed is over 40 mph, leave five seconds for a 40-foot vehicle.

(3) (D) All of the above.

Combination vehicles are likely to skid when not properly handled. You need to be very observant. Look at the mirrors to observe that the vehicle is not skidding. Check your mirrors any time you apply your brakes to be sure the trailer is in the right position. It is not easy to stop an accidental skid or jackknife when the trailer has already gone out of the lane.

(4) (B) More.

The front and rear wheels are often in different positions when you are going around a corner. This is termed cheating or off-tracking. Expect longer trailers to have more cheating than shorter ones.

(5) (A) The service line.

The service air line is also called a signal or control line. It transfers air around the air brake system. The service line pressure is based on the effort you exert on the brake or hand valve. The service line is linked to the relay valves, which facilitate the quicker application of trailer brakes.

(6) (B) It may later skid.

The trailer hand valve is used to test the brakes and should not be engaged in driving because it may later skid. It is better to use the foot brake because it transfers air to every part of the air brake system.

(7) (A) The tractor protection control valve.

The tractor protection control valve maintains the air in the air brake system if there are leaks in the trailer. If the air pressure is around 20 to 45 PSI, the tractor protection valve will close up without manual application.

(8) (C) Emergency line.

The tractor protection control valve causes air to circulate in the air brake system if there are leaks in the trailer. If the air pressure is around 20 to 45 PSI, the tractor protection valve will close up without manual application.

(9) (A) Two.

The two types of air lines are service air lines and emergency air lines.

(10) (D) All of the above.

The service air line is also called a signal or control line. It transfers air around the air brake system. The service line pressure is based on the effort you exert on the brake or hand valve. It is linked to the relay valves, which facilitate the quicker application of trailer brakes.

(11) (B) Supply line.

The emergency air line is also called a supply line. It has two basic functions. The first purpose is to supply air to the air tank. The second is to control the emergency brakes.

(12) (A) The glad hands.

The glad hands or hose couplers are coupling tools that help couple the service and emergency lines in the truck/tractor with the trailer. The couplers come with rubber seals that help prevent air from escaping.

(13) (B) When the emergency and service lines are not connected to the trailer.

Some combination vehicles are designed with dead ends, which are also called dummy couplers. These are connected with the hose when they are idle. This keeps water and dirt particles away from the couplers and the two air lines. Dummy couplers should be used when the emergency and service lines are not connected to the trailer. In the absence of dummy couplers, you can lock both glad hands.

(14) (B) You should open the fifth wheel if it is sliding.

This is not applicable to the supervision of the fifth wheel. When the fifth wheel is sliding, you should lock it up. Inspect the security of your tractor. Check to see if the fifth wheel is well greased to avoid frictional steering problems. Inspect the fifth wheel position and set it in the proper coupling position. Lock up the fifth wheel if it is sliding. Check to see if the kingpin is damaged or out of shape.

(15) (A) Set the trailer directly behind the tractor.

How you position your vehicle matters a lot in the successful coupling of your vehicle. When positioning your vehicle, always make sure the trailer is directly behind the tractor. Use your side mirrors to inspect the position of your tractor.

(16) (C) The tractor may hit the trailer.

The trailer should be reasonably low so that the tractor can bring the trailer up when it is backed below the trailer. A very low trailer may make the tractor hit the trailer and destroy the trailer's nose.

(17) (D) None of the above.

All the answer options are part of the procedure of providing air to trailers. Press the air supply button or switch the valve control from emergency to normal mode so that air can go through the air brake system. You should also look for crossed lines in the air brake system.

(18) (A) Drive slowly.

Drive slowly when backing the tractor under a trailer and stop once there is an interlock between the fifth wheel and the kingpin. When backing up a tractor below a trailer, you should apply the reverse gear, which is at the lowest position.

(19) (A) Increase the gear range.

When the landing gear has become weightless, you should increase the gear range. Keep on lifting the landing gear until it is at the top to avoid it being trapped on railroad tracks.

(20) (D) You should use light surfaces when setting the rig.

This is not true. When setting the rig in the right position, first check the parking area and ensure it is strong enough to take the weight and size of the trailer. Remember that not all surfaces are good supports for the weight of trailers. Establish an alignment between the tractor and the trailer.

(21) (C) To prevent movement.

When uncoupling, chock the trailer wheels to prevent movement of the trailer. This is a very important step because movements can cause disturbances and lead to accidents.

(22) (D) Fixing the air line glad hands with the emergency line.

This statement is false. When disconnecting electrical cables from the air lines, first set apart the air lines and the trailer, then fix the air line glad hands with the dummy couplers, which are at the rear of the vehicle.

(23) (D) All of the above.

When uncoupling the fifth wheel, lift the release handle and place it in the open position. Stand far back from the tractor wheels to prevent injury that may occur if the vehicle moves. Be conscious of the fact that proper care is needed when coupling and uncoupling trailers.

(24) (D) All of the above.

When uncoupling the semitrailer, keep pulling the tractor until the fifth wheel is pulled from the base of the trailer. Pause if the trailer is on the tractor's frame. This will stop the trailer from falling if the landing gear breaks.

(25) (A) Trailer emergency valve.

The tractor protection control valve is also known as the trailer emergency valve. This maintains the air in the air brake system if there is a leak in the trailer or if it breaks. If the air pressure is around 20 to 45 PSI, the tractor protection valve will close up without manual application. This closure prevents air from escaping the air brake system.

22 Answers for Tanker License Test

(1) (D) 1,000.

You will need a tanker license if your tanker is transporting either gases or liquids in a fixed cargo. The cargo tank capacity may be 119 gallons or more, or it can be a portable tank that has a capacity of 1,000 gallons or more.

(2) (D) All of the above.

Do not load, unload or drive until you have inspected or checked the condition of the vehicle to ensure the safety of the liquid or gaseous contents.

(3) (D) All of the above.

Tanker vehicles have great liquid turbulence and an accentuated center of gravity. Therefore, you should not apply the same method of driving that you use with other vehicles. You need to consider the outage, bulkheads, etc.

(4) (B) Heavy.

A greater center of gravity means that a greater part of the weight of the contents of a tanker is raised high and almost off the road. Therefore, the top of the vehicle becomes heavy and is likely to tip.

(5) (C) It does not occur from a wave.

This statement is not true. When the tank is nearly full, expect movement that will lead to a liquid surge. When you are stopping, the liquid will move forward and backward. If a strong wave hits the rear of the tanker, expect the tanker to tilt toward the direction of the wave.

(6) (C) Dangerous.

Liquid surge or overflow is a danger. When the tank is nearly full, expect liquid movement to negatively affect your control of the vehicle.

(7) (D) All of the above.

A bulkhead is used to partition liquid tanks into smaller units. Any time you are loading or unloading, ensure the weight is evenly spread. Avoid overloading the front or the back with excess weight. When you use bulkheads, there will not be an excessive load on one part of the vehicle.

(8) (C) Their holes are for the passage of air.

This statement is not true. Baffled tanks are liquid tanks with bulkheads. They have holes that allow the passage of liquid. You can have better control of the front, back or sideways surge of the liquid when you use baffles.

(9) (B) When transporting food products like milk.

Unbaffled tankers are better used for transporting food products like milk. For sanitation purposes, baffles should not be used to transport food products because it is not easy to clean the tanker.

(10) (B) High.

If you use unbaffled tankers to transport liquid content, expect a very high movement of the tanker. This is because unbaffled tankers do not have any tool to reduce the surging of the liquid.

(11) (A) They do not have the same expansion rate.

Remember that liquids have different expansion rates. Avoid loading a tanker to its brim because liquid expands when warm. Hence, there is the need for outages or room for expansion.

(12) (B) Do not fill up the tank.

If the liquid is very dense (acid), a completely full tank may be beyond the lawful weight limitation. Therefore, if the liquid is dense, do not totally fill up the tank.

(13) (B) Speed of the vehicle.

You will know the liquid is dense by the rate of expansion during transportation, the weight and if it meets the legal weight limitation. Speed does not apply in this situation.

(14) (A) Liquid surge, high center of gravity.

When driving tankers, you have to be very careful to drive smoothly at all times. This is because of the liquid surge and high center of gravity.

(15) (A) Empty tank.

It may be more difficult to stop an empty tank than a full one. Take it slow when you are stopping. Maintain a safe following distance between you and the vehicle in front. Apply stab or controlled braking to stop quickly and prevent a crash.

(16) (D) All of the above.

Do not drive too fast on curved roads. You need to take extra care to avoid bumps or accidents. Do not turn the steering wheel too fast in curves because the vehicle may roll over. If there is no outage, the liquid may overflow.

(17) (D) All of the above.

Too much acceleration or braking may make the vehicle skid. Skidding may make your tanker jackknife. When you drive slowly, you will be able to effectively avoid skids. Always remember that tankers have a high center of gravity and are likely to skid if care is not taken.

(18) (D) All of the above.

To thoroughly inspect for leaks, check the body or outer part of the tank, the valves, all the connections and pipes, the vents and manhole covers. Ensure there are gaskets to keep the covers well closed. Also ensure the vents are free from particles or obstructions for efficient functioning.

(19) (D) All of the above.

When inspecting the valves, make sure there is an intake, discharge and shut off. Fix the valves in the most appropriate places before you start loading, unloading or driving the tanker. Also, inspect all the connections and pipes. Inspect the vents and manhole covers. Ensure there are gaskets to keep the covers well closed.

(20) (C) Properly closed.

Ensure there are gaskets to keep the manhole covers properly closed.

(21) (C) Refrain from transporting packages.

You should refrain from driving tank vehicles or transporting packages whose valves and manhole covers are not closed. This is very dangerous. Always perform a practical inspection of the necessary equipment. Test the equipment before you drive.

(22) (A) Wait for enough of a gap so you can merge.

Do not be in a hurry to cross traffic. First, confirm if the space is long enough for your tanker vehicle. Before changing lanes while driving, confirm that no vehicle is next to you.

(23) (D) Move the gear to the right downshift.

In curves, slow down and downshift to the right-hand side when approaching a curve. It helps keep the vehicle stable. With your gear in this position, you can also easily speed up after driving through the curve.

(24) (A) Where backing up is easy.

You should park where backing up is easy because it is difficult. Backing up should be avoided as much as possible. Look in every direction when parking. Use the mirrors to watch your sides and rear view. There should be enough space to move forward when leaving the parking lot because sometimes pulling out can be difficult, especially if the space is too tight.

(25) (A) Someone might mistakenly drive the vehicle and crush whoever is under it.

The reason for removing the keys from the ignition or starter before you start a pre-trip inspection is to prevent a situation where someone moves the truck while you are still under the vehicle.

23 Answers for Passenger Transport Test

(1) (C) A priority.
Safety is a priority when driving a passenger vehicle. You must ensure every object and person you are driving is safe before loading and embarking on a trip.

(2) (A) Read through it before driving.
Go through the inspection report that the previous driver made to see what is lacking. Refuse to sign it if what needs to be repaired or replaced has not been done.

(3) (D) All of the above.
Before you drive, test the parking brake; the service brakes, which may include air hose couplings for vehicles with a semitrailer or trailer; the reflectors and the lights; the steering system; the horn; the tires; the wipers; the mirrors; coupling tools; the wheels and the emergency kits.

(4) (C) Is very important and should be inspected.
The wiper is very important and should always be inspected.

(5) (A) The interior.
The railing is an interior part of the vehicle.

(6) (C) Before loading.
Before you load and drive your vehicle, you should check the floor covering and ascertain that it is safe before anyone enters.

(7) (B) You can partially close up the roof hatches.
If you need fresh air, you may partially close up the emergency roof hatch.

(8) (A) Spare parts.
Ensure there are spare parts, such as tires and electrical fuses. These components are vital in emergencies such as fires.

(9) (C) Replace them with new ones immediately.
Seat belts are essential components of the vehicle. Always make use of yours and advise your passengers to use theirs. If there is wear and tear on seat belts, replace them.

(10) (D) All of the above.
Obstructions may make passengers stumble and get injured. Remove baggage from the aisle or doorways. Show riders where to keep their baggage. Inform riders that they can exit through the window or the closest door in an emergency.

(11) (D) All of the above.

Transporting hazardous materials requires extra care. You are not allowed to transport hazardous materials if you do not have a hazmat license or endorsement. Ensure you are aware of what every cargo or baggage contains. Never forget that hazardous materials are threats to the health, property and safety of the people who are around. Mark and label these as hazmats.

(12) (D) All of the above.
Buses are allowed to drive small-arms ammunition that is ORM-D, emergency supplies for a hospital and drugs. Mark those hazmats that you are allowed to transport.

(13) (D) All of the above.
Hazardous materials are threats to the health, property and safety of the people who are around. Therefore, you are not allowed to transport certain hazardous materials on passenger vehicles. These include radioactive materials near passenger seats, explosives near seats, solid poisons that weigh more than 100 pounds, etc.

(14) (A) 500 pounds.
You are not allowed to transport hazardous materials that have a total weight of more than 500 pounds.

(15) (D) None of the above.
The standee line shows passengers where they can and cannot stand.

(16) (A) No passenger should stand behind the driver's seat.
This statement is true.

(17) (D) All of the above.
None of these statements are true. When you are almost at a bus stop or final destination, announce the destination, state why you are about to stop, announce the time of departure before and/or after every stop and announce the number of the vehicle.

(18) (D) All of the above.
Before you begin a trip, clearly state the rules of no smoking, drinking, sexual intercourse, loud music and other rules. Be sure that passengers understand each of the rules and answer any questions before you begin the trip. Also ensure everyone is securely seated before driving.

(19) (C) The mirrors.
Use the mirrors to supervise what the passengers are doing while you are driving. You may have to caution offenders and remind them of the rules. Correct those whose heads or arms are outside the windows. You must be vigilant.

(20) (C) Carefully handle the situation.

If you encounter a belligerent rider, carefully handle the situation to ensure everyone is safe. Do not be in a hurry to drop off the belligerent rider until it is safe to do so. You may have to wait until the next stop before you have the person leave the vehicle.

(21) (A) Traffic signs are enough to guide drivers.

This statement is false. Most bus crashes occur at intersections. Do not depend on traffic signs alone. Be observant enough to notice oncoming vehicles, poles, or objects in the way. Sometimes you may not be the driver at fault, but you still need to prepare for the unexpected incidents that reckless drivers may cause.

(22) (C) Slippery roads, speed.

Slippery roads and speed are the greatest enemies to safety. They often result in the loss of lives and properties. Most roads are designed with an acceptable speed limit. However, the speed limit for a car may not be what is right for a bus. So you must be careful with the speed of your vehicle.

(23) (D) All of the above.

When stopping at railroad-highway crossings, you should apply the brake when your bus is about 15 to 50 feet from the railroad crossings. Watch out for trains by listening and looking. If possible, take a step out of your seat to be sure if there are incoming trains or not. Do not be too quick to jump into a lane after the train has passed.

(24) (B) Wait to be sure a space is clear.

When stopping at railroad-highway crossings, you should make sure you wait to be sure a space is clear before proceeding with your journey.

(25) (D) All of the above.

Do not refuel a bus if there are riders in the vehicle, except if it is very necessary. Be focused while driving. Do not be carried away with chitchat or anything that takes your mind off the road or from your passengers. Except when the safety of your passengers is threatened, never push or tow a faulty bus when the passengers are onboard. And if the area is not safe, push the bus to the nearest safe zone and unload the passengers so that you can continue pushing or towing.

24 Answers for School Bus Endorsement Test

(1) (C) Use your own disciplinary measures for the situation.

This statement is false. When managing students, you should always observe the school rules and regulations on discipline before, during and after transit. Stop the bus as soon as you find a safe parking area. Turn off the vehicle and take your key as you step out of it. Remind the offender(s) of the rules and regulations.

(2) (D) All of the above.

Students should be reminded of the rules and properly monitored using the mirror. Your anger should not be expressed facially or in your tone, but you should remain firm. Settle disputes according to the disciplinary code of the school.

(3) (D) All of the above.

On no occasion should you remove a student from the bus until you have arrived at the official bus stop. You may call the school administration or the police and inform them so that they can come and pick up the student if the offense is very serious. Just remember to comply with protocols on discipline.

(4) (D) All of the above.

When approaching a designated bus stop, you must be very cautious with your speed. This area demands vigilance and skill. Without caution, you may crash into other vehicles or pedestrians. You will have to use the mirrors, the lights, the signal arm and the crossing control arm. Watch both sides of your vehicle.

(5) (D) All of the above.

When arriving at a designated destination, keep checking the mirrors. Stop the school bus 10 feet from the bus stop. Do not set the transmission in neutral or park. Wait for total clearance. You can open the service door to switch on the red lights if the traffic is a bit far from the school bus.

(6) (C) When approaching the bus stop.

When approaching the school bus stop, you should switch on the warning lights a few seconds (5 to 10) or 100 to 500 feet ahead. Switch on the right-side indicator for three to five seconds when you are about to pull over. Stay very close to the side of the driving lane. Stop the school bus 10 feet from the bus stop.

(7) (B) 10.

Stop the school bus 10 feet from the bus stop. Doing this will make the students walk toward the bus. It will help prevent accidents and also will allow you to count the students.

(8) (A) Very close to the side of the driving lane.

When parking, you need to stay very close to the side of the driving lane to prevent accidents. This will enable the students to easily gain access to the bus.

(9) (D) All of the above.

Wait for traffic to totally clear before you open the door and ask the students to start coming in. Stay very close to the side of the lane. Set the transmission in neutral.

(10) (A) When the driver signals.

No students should enter the bus until the driver has signaled them to come in.

(11) (B) Check around and beneath the vehicle.

Always remember to take a count of the students. If the number of students on board does not tally with the expected number, ask for the whereabouts of the missing student(s). If you are not satisfied with the answer, safely park the bus and check around and beneath the bus.

(12) (B) Turn on the dome light.

Turn on the dome light if the loading area is dark. When the students are entering the bus, ask them not to rush in. Instead, they should be in a single-file line and use the handrails.

(13) (A) Walk about 45 feet.

This statement is false. If the students will cross the road, step down from the bus and walk about 10 feet from the bus to where you can see the students crossing. Stay at the right side of the road, where you can clearly see the feet of the students. Look in all directions and ensure the road is clear for crossing.

(14) (D) All of the above.

When it seems like you have fully unloaded at your final destination, check inside the bus to be sure that there are not any students who have not unloaded for any reason. After parking the bus and unloading the students, count all the students and be sure none is missing. If any are missing, check around to find them.

(15) (A) Let go of the parking brake.

After you have accounted for each student, prepare to leave by shutting the door, putting on the seat belt, turning on the engine, running the transmission, letting go of the parking brake, switching off the alternating flashers, switching on the left turn signal and looking in the mirrors again. Then drive away.

(16) (B) Danger zone.

A school bus danger zone is an area where schoolchildren are likely to receive the greatest injury from their bus or another bus.

(17) (A) 30.

The danger zones of your bus may be 30 feet away from the front bumper, with the first 10 feet being often more dangerous.

(18) (A) They are optional components of your vehicle.

This is false. Mirrors are not optional vehicle components. To avoid accidents when driving, you need to properly adjust the mirrors, which are very useful for looking out for traffic, objects and the welfare of the students. Ensure all the mirrors give you the right visual field.

(19) (C) 50 to 150 feet.

The blind spot is the side of the road that the mirrors cannot capture. A blind spot is directly in front of and under each flat mirror and behind the rear bumper. The distance of the blind spot is about 50 to 150 feet; sometimes it goes as far as 400 feet based on the size of the school bus.

(20) (B) The length of 4 buses.

When you properly adjust the outside flat mirrors, it will help you have a view of approximately 200 feet or the length of four buses at the back of your bus; the sides of the bus and the back tires where they make contact with the ground.

(21) (B) The outside convex mirrors.

The outside convex mirrors are directly under the flat mirrors outside your vehicle. You use them to have a panoramic view of what is happening on the left and right sides of your environment outside. Like the flat mirrors, they give you a view of the students who are on the side, traffic and other ongoing activities. These mirrors will not give you the exact distance and size of objects.

(22) (B) The outside convex mirrors.

Properly adjusting the outside convex mirrors will help you have a view of every side of the bus, the front part of the back tires where they have contact with the ground and the traffic lane outside.

(23) (D) All of the above.

Properly adjusting the outside crossover mirrors will help you have a view of every zone ahead of the bus, starting from the bumper at the front to where you can have a direct vision. The combination of mirror visuals and your eyes is best. Both should overlap and not serve as substitutes for one another. The mirror also helps you view all the front tires where they do not make contact with the ground.

(24) (D) All of the above.

Expect to have a blind spot behind the driver's seat and the back of the bumper that can go as far as 400 feet behind the bus. To have a view of this blind spot, use the side mirrors. This way you can monitor the incoming traffic around the blind spot area.

(25) (D) None of the above.

None of the answers are right. There are two types of crossings: passive crossings and active crossings. Passive crossings do not have a traffic control machine. You have to stop and apply the normal traffic procedures. Active crossings come with a traffic control machine, which regulates traffic. These active crossing devices are red lights that may or may not have bells.

25 General Knowledge Test

1. A Class A license is for drivers who operate which types of vehicles?
A. Two or more vehicles with a combined weight of less than 25,000 pounds
B. Two or more vehicles with a combined weight of over 25,000 pounds
C. Two or more vehicles with a combined weight of less than 26,000 pounds
D. Two or more vehicles with a combined weight of over 26,000 pounds

2. A Class A driver's license can be used to drive a wide range of vehicles except:
A. Tractor-trailers
B. Tank vehicles
C. SUVs
D. Flatbeds

3. A P Endorsement allows the holder to drive a passenger vehicle that carries a minimum of ... passengers.
A. 15 passengers
B. 20 passengers
C. 16 passengers
D. 10 passengers

4. What is the TSA?
A. The Transport Security Administration
B. The Transport Safety and Security Administration
C. The Transportation Safety Administration
D. The Transportation Safety and Security Administration

5. What class of vehicle does the N Endorsement cover?
A. Tanks
B. Flatbeds and SUVs
C. School buses
D. Passenger buses

6. If you want to be a school bus driver, you need the ...
A. P Endorsement
B. K Endorsement
C. S Endorsement
D. Z Endorsement

7. Which of these states doesn't allow drivers to operate triple vehicles?
A. North Dakota
B. West Virginia
C. Delaware
D. New Hampshire

8. The ... license is issued to drivers who operate commercial vehicles with a combined weight over 26,000 pounds.
A. Class A
B. Class B
C. Class C
D. Class D

9. With the appropriate endorsement, you can drive ... vehicles with a Class B license.
A. Class D
B. Class A
C. Class F
D. Class C

10. Dump trucks, segmented buses, large passenger buses and straight trucks are some of the vehicles that can be driven with a ... license.
A. Class D.
B. Class B.
C. Class A.
D. Class C.

11. The required license class for transporting hazardous materials is ...
A. Class C.
B. Class D.
C. Class B.
D. Class A.

12. The Commercial Drivers' License Test is divided into ...
A. Five sections.
B. Eight sections.
C. Ten sections.
D. Four sections.

13. These test sections include all of the following except ...
A. Combination vehicles.
B. General knowledge.
C. Hazardous materials.
D. Private cars.

14. A driver is deemed qualified to receive a Commercial Driver's License if he/she gets a minimum score of ... on the test.
A. 70%.
B. 80%.
C. 75%.
D. 90%.

15. Which of the following candidates passed the CDL test?
A. A candidate who scored 35 out of 50.
B. A candidate who got 45 answers right.
C. A candidate who scored between 35 and 49 out of 50.
D. None of the candidates.

16. The examination is ... long.
A. 50 minutes.
B. 45 minutes.
C. 60 minutes.
D. 90 minutes.

17. You must present the following documents in order to be admitted to the testing center:
A. Proof of identity and proof of social security number.
B. Proof of social security number and utility bills.
C. Proof of identity and two guarantors.
D. Proof of citizenship and credit history.

18. Proof of identity includes all of the following except:
A. Credit history.
B. A government-issued identity card.
C. U.S. military discharge papers.
D. A birth certificate.

19. Which of these types of documents are acceptable?
A. Photocopies and original copies of your documents.
B. Duplicate documents and photocopies.
C. Original documents only.
D. Duplicate documents and original copies of your document.

20. You can prove your citizenship or permanent residency with the following:
A. Certificate of citizenship and birth certificate.
B. Permanent residency papers and utility bills.
C. US passport and work permit.
D. All of the above.

21. Other acceptable proof of residency documents are stated in the income tax return otherwise known as the...
A. IRS W-2 form.
B. IRS W-1 form.
C. IRS P-1 form.
D. IRS P-2 form.

22. The order of name arrangement is immaterial when filling out the Commercial Driver's License application form.
A. Yes.
B. No.
C. The DMV must decide.
D. It depends on several factors.

23. A Commercial Learner's Permit holder must wait for a minimum of...
A. Two months.
B. Two weeks.
C. Three days.
D. Ten days.

24. The minimum acceptable age to obtain a driver's license in most states is...
A. 18 years.
B. 21 years.
C. 20 years.
D. 15 years.

25. Applicants between... are only allowed to drive on the interstate.
A. 19 and 20.
B. 19 and 21.
C. 18 and 20.
D. 15 and 19.

26. The interstate permit expires...
A. When you are 22 years old.
B. When you are 30 years old.
C. When you are 21 years old.
D. Not for as long as you are in the driving business.

27. The minimum age requirement for interstate driving is
A. 35 years old.
B. 28 years old.
C. 21 years old.
D. There is no minimum age.

28. All of the following may disqualify you from obtaining a Commercial Driver's License except:
A. Inability to speak or read English fluently.
B. Failure to produce proof of identity on request.
C. Providing an expired government-issued identity card.
D. None of the above.

29. A disqualified driver can have the disqualification overturned after how many years?
A. Five years.
B. Two years.
C. Ten years.
D. The decision cannot be overturned.

30. Drivers who wish to operate commercial motor vehicles must have... vision in each eye.
A. 20/40
B. 30/50
C. 40/20
D. 50/50.

31. The acceptable value for horizontal vision is...
A. 130 degrees or higher
B. 140 degrees or higher.
C. 150 degrees or higher.
D. 100 degrees or higher.

32. You may be disqualified from taking the test if you are found with the following items at the testing center:
A. Written notes or other materials.
B. Electronic devices such as cell phones.
C. Personal effects.
D. All of the above.

33. Driving under the influence of alcohol or drugs may result in...
A. A year-long ban from applying for a CDL license.
B. A five-year ban from applying for the CDL.
C. A fine of $500.
D. Permanent disqualification from applying for the license.

34. The acceptable blood alcohol content acceptable for driving a commercial vehicle is...
A. Below 0.10.
B. Below 0.70.
C. Below 0.04.
D. Below 0.05.

35. If your blood alcohol content is above the acceptable value, you may be prohibited from driving for...
A. 30 to 60 days.
B. 60 days to 180 days.
C. 30 to 45 days.
D. 60 days to life.

36. What is the penalty for committing a felony with a commercial vehicle?
A. You may be banned for life.
B. You will be banned for a minimum of five years.
C. You will be banned for at least three years.
D. You will be banned for a year.

37. If you are convicted of manufacturing or distributing drugs illegally with a commercial motor vehicle, the penalty is:

A. A lifetime ban.

B. A five-year ban.

C. A six-month ban.

D. The duration of the ban is dependent on various factors.

38. Failing to stop at an accident scene that involves your vehicle, especially if the accident leads to a loss of life or injury, may incur:

A. A $100 fine.

B. A $500 fine.

C. A lifetime driving ban.

D. A year-long driving ban.

39. The T endorsement allows you to drive the following vehicles:

A. Double or triple trailers.

B. Double trailers only.

C. Triple trailers only.

D. None of the above.

40. What should be your first action if your truck starts to slide on ice and you are about to lose control?

A. Apply the parking brake immediately

B. Turn the steering wheel in the opposite direction of the slide

C. Slowly release the accelerator

D. Swiftly change the gear to reverse

41. Why should the parking brake be avoided when a truck is sliding on ice?

A. It can cause the truck to slide further

B. It can cause the truck to flip over

C. It can cause the tires to freeze

D. It can cause the brake pads to wear out faster

42. When is it considered safe to use the parking brake on a truck sliding on ice?

A. When you want to turn sharply

B. When the truck is at a complete stop

C. When you want to slow down quickly

D. When the truck is moving downhill

43. How should you regain control of your truck if it begins to slide on ice?
A. Quickly apply the parking brake
B. Steer in the direction of the skid and lightly pump your service brakes
C. Increase the speed to gain momentum
D. Make a sharp turn in the opposite direction of the slide

44. What can happen if you apply the parking brake abruptly while your truck is sliding on ice?
A. The vehicle will stop immediately
B. The vehicle may start to jackknife
C. The vehicle will regain traction
D. The vehicle will speed up

45. What is CLP?
A. Community License Producers.
B. Commercial Learning Permit.
C. Commercial Learner's Permit.
D. Commercial Learner Permit.

46. What is the first action you should take if you experience a blowout while driving your truck?
A. Steer hard to the right
B. Brake immediately
C. Maintain your speed and steering control
D. Accelerate to avoid losing control

47. What is one common sign that a truck tire has gone flat while driving?
A. The truck starts to accelerate faster
B. There is an unusual vibration or thumping noise
C. The truck fuel consumption decreases
D. The truck steers to the left or right with no input from the driver

48. In case of a flat tire, why is it important to inspect the other tires on the same axle?
A. To check if they are also flat
B. To check if they need to be replaced as well
C. To check if they have been affected by the heat or debris from the blown tire
D. All of the above

49. After changing a flat tire on your truck, why should you immediately check the tire pressure of the new tire?

A. To ensure it is inflated to the manufacturer's recommended pressure

B. To make sure it has not been punctured during the change

C. To check if it is compatible with the truck

D. To test if the tire pressure gauge is working properly

50. If your truck's tire blows out while you're driving, what should you avoid doing?

A. Avoid braking until the truck has slowed down

B. Avoid changing lanes until you have control of the vehicle

C. Avoid turning on your hazard lights to warn other drivers

D. Avoid steering against the pull of the vehicle

26 Air Brakes Test

1. Air brakes are otherwise known as a...
A. Compressed air brake system.
B. Compressed air braking system.
C. Compressed pressure air brake system.
D. Compressed pressure air braking system.

2. The brake system is powered by...
A. Compressed air.
B. Pressurized air.
C. All of the above.
D. None of the above.

3. The part of the brake system that is responsible for stopping the vehicle is the...
A. Brake pad.
B. Brake pressure.
C. Brake shoe.
D. Brake stopper.

4. The primary difference between vehicles with air brakes and those with hydraulic fluid is that...
A. Air brakes are not affected by leaks while hydraulic brakes are.
B. Hydraulic brakes are not affected by leaks while air brakes are.
C. There is no significant difference between the two.
D. Air brakes are more expensive than hydraulic brakes.

5. Reliability is one of the major factors behind the rising popularity of...
A. Hydraulic fluid-based brake systems.
B. Air brake systems.
C. Both brake systems.
D. Neither brake system.

6. ... developed air brakes for use in railways.
A. George Washington.
B. George Barbados.
C. George Westinghouse.
D. George Stones.

7. The air brake was patented on...
A. March 10, 1880.
B. March 15, 1890.
C. March 5, 1872.
D. March 20, 1875.

8. ... are the most common air brake systems.
A. Common brakes.
B. Foundation brakes.
C. Superb brakes.
D. Permanent brakes.

9. The components that hold enough air for brakes to use repeatedly are the...
A. Air containers.
B. Air reservoir tanks.
C. Air storage tanks.
D. Air preservation tanks.

10. This component pumps air into the reservoir tank.
A. Air pump.
B. Air release mechanism.
C. Air compressor.
D. Air distributor.

11. What function does the slack adjuster perform?
A. It connects the push rod and the brake through its arm.
B. It connects the push rod and the brake chambers.
C. It connects the air compressor and the push rod.
D. It connects the brake chambers and the brake.

12. The brake pedal is otherwise known as...
A. The foot valve.
B. The brake engager.
C. The brake depressor.
D. The brake applier.

13. The... reduces the risk of the air brake valves freezing up.
A. Air evaporator.
B. Alcohol evaporator.
C. Ice breaker.
D. None of the above.

14. Two major contaminants of air tank drains are... and...
A. Water and oil.
B. Water and dirt.
C. Oil and other particles.
D. All of the above.

15. This type of air tank comes equipped with electric heating devices.
A. Electrical air tanks.
B. Manual air tanks.
C. Automatic air tanks.
D. Hybrid air tanks.

16. Spring brakes serve what purpose in the air brake?
A. They provide the mechanical force that prevents accidental leakage of air pressure.
B. They are used for holding the brake pedals together.
C. They establish a connection between the alcohol evaporator and the return spring.
D. They protect the brake from excessive pressure.

17. Air brakes are fully operational when the air pressure drops to...
A. Between 30 and 40 psi.
B. Between 15 and 20 psi.
C. Between 20 and 30 psi.
D. Between 30 and 50 psi.

18. What should you do when the buzzer and low air pressure warning light flashes?
A. Stop the vehicle while you can still control the brake.
B. Drive for as long as possible.
C. Drive until you can stop the warning signals.
D. Do nothing.

19. Braking power is dependent on...
A. How the brakes are adjusted.
B. The air pressure in the brake system.
C. The parking brake controls.
D. The types of drain valves in the brake system.

20. Which valve doesn't need assistance to expel water and oil?

A. Manual valve.

B. Hybrid valve.

C. Automatic valve.

D. All of the above.

21. The major cause of brake failure in cold weather is...

A. Bad roads.

B. The water in the tank may freeze.

C. The brake pedals cracking due to cold water.

D. The alcohol to prevent freezing was not circulated properly in the system.

22. Which part of the air brake separates the brakes and pushes them against the brake drum?

A. The brake S-Cam.

B. The safety valve.

C. The air compressor governor.

D. The slack adjuster.

23. A steel rod in the air brake that shares some similarities with the piston connecting the brake chamber and the slack adjuster is...

A. The brake rod.

B. The connecting rod.

C. The push rod.

D. The slacking rod.

24. A lined steel mechanism that produces friction when it comes in contact with the brake drum is the...

A. Brake lining.

B. Brake compartment.

C. Brake shoe.

D. Brake connector.

25. Why is it important to drain the air tank regularly?

A. It prevents brake malfunctioning by preventing contaminants from affecting the brakes.

B. It provides room for more air in the air tank.

C. It is a preventive measure against corrosion.

D. It purifies the stored air.

27 Doubles and Triples Test

1. What are doubles and triples?
A. Long vehicles with two or more trailers attached to the truck's tractor.
B. Long vehicles with a trailer or two attached to the truck's tractor.
C. Long vehicles with two or more truck tractors.
D. Long vehicles with more than two truck tractors.

2. Doubles and triples are also known as...
A. Longer Commercial Vehicles.
B. Longer Combination Vehicles.
C. Large Commercial Vehicles.
D. Large Combination Vehicles.

3. Do states have common restrictions on triples?
A. Yes.
B. No.
C. It depends on the type of triples.
D. It depends on the type of license a driver holds.

4. What does an assembly procedure entail?
A. It is a test of your ability to assemble a vehicle from scratch.
B. It is a test of your ability to fix whatever you uncouple.
C. It is a test of your ability to operate vehicles driven directly from assembly plants.
D. It is a test of whether you meet vehicle assembling requirements or not.

5. What is a test of your ability to operate a vehicle irrespective of potential challenges?
A. Handling procedure.
B. Handling requirements.
C. Handling.
D. Handling problem solutions.

6. To take the Doubles and Triples test you must pay a:
A. Licensing fee.
B. Testing fee.
C. None of the above.
D. All of the above.

7. Your choice of test location is dependent on several factors that include...
A. Your state of residence.
B. The type of double or triple you drive.
C. Your license type.
D. Your learner's permit.

8. One of the most common problems associated with pulling doubles or triples is...
A. Their weight.
B. They are too long.
C. They are less stable than other vehicles.
D. All of the above.

9. What is a preventive measure against rolling over in doubles or triples?
A. Carrying fewer loads.
B. Driving slowly.
C. Steering gently.
D. Both B and C.

10. The crack-the-whip effect usually occurs when...
A. A trailer driver applies the brakes suddenly.
B. A trailer driver changes lane quickly.
C. The trailer is overloaded.
D. The driver increases the driving speed suddenly.

11. It is advisable to leave space between you and the vehicle ahead of you. Why?
A. To enable your vehicle to turn suddenly.
B. To make it easier for you to overtake the vehicle ahead.
C. To allow smaller vehicles to overtake your trailer easily.
D. To make it easier to maneuver your vehicle.

12. ... and ... are problems common to doubles and triples when mountain driving.
A. Skidding and loss of traction.
B. Loss of traction and jackknifing.
C. Skidding and careless overtaking.
D. All of the above.

13. One of several factors you must consider when parking is...
A. Ease of driving out of the parking lot.
B. Ease of driving into the parking lot.
C. Ease of using the brakes when necessary.
D. Ease of managing bad weather.

14. An important step towards ensuring your trailer's emergency brakes function properly is...
A. Securing the trailer.
B. Oiling the emergency brakes.
C. Connecting the emergency brakes with a fifth-wheel dolly.
D. Chocking the wheels.

15. The semitrailer with the heaviest load should be connected...
A. In front of the tractor.
B. Behind the tractor.
C. Behind other semitrailers.
D. Wherever you are comfortable.

16. Where is the most appropriate location for the converter dolly?
A. In front of the main first trailer.
B. In front of the tractor.
C. Beside the rear trailer.
D. In front of the rear trailer.

17. How do you connect the first semitrailer to the converter dolly?
A. By locking the pintle hood and securing the converter gear support.
B. By directly locking the pintle hood and converter dolly.
C. By connecting them through a connector hook.
D. Using a specially designed converter connector.

18. During a trailer coupling, there shouldn't be spaces between the... and...
A. Lower and upper fifth wheel.
B. Converter dolly and kingpin.
C. Safety chains and light cords.
D. Air tank petcock and first trailer.

19. How do you charge the brakes?
A. By applying more pressure to the brake pad.
B. By reducing pressure to the brake pad.
C. By pushing the air supply knob in.
D. By shutting the valves off.

20. What are the most important benefits of inspecting your vehicle regularly?
A. It makes the vehicle safer to drive.
B. It saves time.
C. It satisfies your conscience.
D. All of the above.

21. Where can you take the doubles and triples test?
A. At a third-party site.
B. At the local Department of Motor Vehicles.
C. All of the above.
D. None of the above.

22. What does LCV stand for?
A. Longer Commercial Vehicle.
B. Longer Communication Vehicle.
C. Larger Combination Vehicles.
D. Longer Combination Vehicles.

23. The doubles and triples test covers all of the following except:
A. Assembly procedures.
B. Potential problems.
C. Handling.
D. Purchasing and supply.

24. All of the following factors make operating a double or triple trailer more challenging except:
A. The cost of purchase.
B. Its sheer size.
C. Its outstanding weight.
D. The incredible length.

25. While uncoupling the rear trailer, you must take the following precautions:
A. Park the rig on level and firm ground.
B. Park the rig in a straight line.
C. Remove some of its load.
D. All of the above.

28 Hazmat Test

1. Hazmat means...
A. Hazardous matters.
B. Hazardous materials.
C. Hazardous metallic substances.
D. None of the above.

2. Flammable solids such as...
A. Matches and fuses.
B. Gasoline and ammonia.
C. Acetone and chlorine.
D. Fuses and sulfuric acid.

3. Two major factors that determine the hazardous nature of a substance are...
A. Characteristics and shipper's decision.
B. Shipper's decision and shipping technique.
C. Characteristics and nature.
D. Weight and characteristics.

4. The hazmat test should be taken in...
A. French and English.
B. English.
C. French.
D. In a driver's preferred language.

5. Which of the following shows that a material is hazardous?
A. It's placarded with an X or RQ.
B. A diamond label.
C. All of the above.
D. None of the above.

6. The following usually are classified as hazardous materials:
A. Fireworks and pest control.
B. Chemical supplies and pencils.
C. Agricultural materials and books.
D. Paints.

7. These endorsements are associated with transporting hazmat.

A. H or X.

B. G or P.

C. K or L.

D. D and H.

8. To qualify for a Commercial Driver's License to transport hazardous materials, you must:

A. Take a security threat assessment test and be aged 21 and above.

B. Take part in a security threat assessment test and be aged 18 and above.

C. Take part in a security threat assessment test and be between 18 and 21.

D. Take part in a security threat assessment test and be above 30 years old.

9. For your CDL renewal, you are expected to provide up to a... history of the licenses you have held in the past.

A. 15-year.

B. 8-year.

C. 10-year.

D. 18-year.

10. To qualify for the CDL Hazmat license, you must pass the...

A. TSA Hazmat Security Assessment.

B. BSA Security Threat Assessment.

C. TSA Security Threat Assessment.

D. BSA Hazmat Security Threat Assessment.

11. The TSA test rule was established on...

A. May 10, 1890.

B. May 5, 1980.

C. May 5, 2003.

D. May 10, 2008.

12. You have... to complete the TSA assessment after submitting your fingerprints.

A. 20 days.

B. 15 days.

C. 30 days.

D. 45 days.

13. The assessment completion date can be extended if...
A. You have eligibility issues.
B. You apply for an extension.
C. There are insufficient materials for the test.
D. The examiner deems it fit to extend the date.

14. You can be permanently barred from getting the hazardous materials endorsement if...
A. You are late to the exam center.
B. You fail to complete the test.
C. You fail the test three times in a row.
D. You have a criminal record.

15. Which of these crimes incur automatic disqualification from the hazmat endorsement?
A. Espionage and sedition.
B. Treason and theft.
C. Public fighting and excessive speeding.
D. All of the above.

16. You are forbidden from taking this test if you are...
A. A fugitive.
B. A feminist.
C. An animal rights activist.
D. A social activist.

17. Other crimes that may prevent you from taking the test include all of the following except:
A. Arson and kidnapping.
B. Environmental damage and disruption of the transportation system.
C. Tax evasion and hostage taking.
D. None of the above.

18. You can challenge a ban if you have proof of the following:
A. US citizenship or permanent residency.
B. Mistaken identity and reversed conviction.
C. International driver's license and international passport.
D. Years of experience in the military or public health sector.

19. You have the right to apply for a waiver if you have proof of:
A. Acknowledgment of conviction based on reason of insanity.
B. Acknowledgment of conviction based on a strong personal conviction.
C. Accidental murder or manslaughter.
D. A and C.

20. The TSA will respond to your waiver application within...
A. 10 days of the receipt of your waiver application.
B. 5 days of the receipt of your waiver application.
C. 15 days of the receipt of your waiver application.
D. 30 days of the receipt of your waiver application.

21. Which of these crimes doesn't mean disqualification?
A. Roadside infractions.
B. Placarding violations.
C. Bribery.
D. A and B.

22. The Security Threat Assessment is conducted by the...
A. Transportation Security Arbitrators.
B. Transportation Security Amendment Body.
C. Transportation Security Association.
D. Transportation Security Administration.

23. Which of the following criteria must you meet to be eligible for the hazmat endorsement?
A. State government criteria only.
B. Federal government criteria only.
C. State and federal government criteria.
D. State criteria is mandatory while federal criteria is optional.

24. When inspecting the tanks used for transporting hazardous materials, check for the following:
A. Diamond labels on the premises.
B. Placard labels on the premises.
C. None of the listed signs.
D. All of the listed signs.

25. The safest transportation vessel for hazardous materials is...
A. Any type of commercial vehicle.
B. Planes and ships.
C. Heavy commercial vehicles.
D. All of the above.

29 Combination Vehicles Test

1. To handle a wide range of heavy commercial vehicles, you must pass the ________.
A. Heavy vehicles test
B. Combination vehicle test
C. Heavy commercial vehicles test
D. All of the above

2. What attributes differentiate combination vehicles from other commercial vehicles?
A. They are heavier and longer
B. They are more expensive and longer
C. They require different licenses to drive
D. They break down easily

3. Combination vehicle drivers must ________.
A. Have excellent knowledge of the roads and their vehicle
B. Have excellent driving experience and top-notch knowledge of their vehicle
C. Be more experienced than drivers of other vehicles
D. Have five years of driving experience

4. Which test comes first?
A. The driving test
B. The written test
C. The knowledge test
D. The ability test

5. A combination of vehicles is otherwise known as ________.
A. A semitruck
B. A semitrailer
C. A semitruck-trailer
D. A semi-tanker

6. Combination vehicles can easily bend in the middle thanks to a ________.
A. Fifth-wheel dolly
B. Fifth-wheel converter
C. Fifth-wheel bender
D. Fifth-wheel combination connector

7. Which of the following is easier to stop?
A. An empty truck
B. A fully loaded truck
C. A half-full truck
D. A new truck

8. What is jackknifing?
A. A condition in which a fully-loaded combination vehicle stops abruptly
B. A condition in which the vehicle bends into a V shape while skidding uncontrollably
C. The joining of vehicles with a converter dolly
D. Easily stopping an empty truck

9. ________ is the primary cause of jackknifing.
A. Loss of traction
B. Loss of direction
C. Speeding
D. Careless braking at corners

10. Jackknifing is common to what types of combination vehicles?
A. Empty trailers
B. Fully loaded trailers
C. Trailers with badly distributed loads
D. Both A and B

11. What must you avoid when driving on a curve?
A. Decelerating and braking sharply
B. Looking at the vehicles behind
C. Honking your horn
D. All of the above

12. One way to guard against jackknifing is to ________.
A. Spread your braking over a distance
B. Break at intervals
C. Apply the brakes at your convenience
D. Honk the horn before applying the brakes

13. A prelude to jackknifing is ________.
A. Skidding
B. Braking
C. Horn honking
D. Overtaking

14. You can avoid a wide range of braking-related problems if ________.
A. You apply the brakes regularly
B. You drive with full concentration on your brakes
C. You pay attention to your environment
D. You don't panic when you apply the brakes suddenly

15. What is cheating?
A. A situation in which both the rear and front wheels take different paths
B. When the brakes fail to function properly after applying them
C. When you are forced to give way to a reckless driver who overtakes you carelessly
D. A situation that tests your driving skills when negotiating a bend

16. You can control cheating by ________.
A. Ensuring no vehicle overtakes you carelessly
B. Ensuring that your brakes are in good condition
C. Focusing on the front wheels
D. None of the above

17. A preventive measure against getting stuck on railroad tracks is ________.
A. Not shifting gears while crossing railroad tracks
B. Shifting gears midway while crossing railroad tracks
C. Stopping your vehicle when necessary on the railroad tracks
D. Paying attention to your environment while crossing railroad tracks

18. What role does a DOT placard play while crossing a public railroad?
A. It contains a number you can call for immediate assistance during emergencies
B. It contains driving information such as the speed limit
C. It contains information about the nearest gas station
D. It provides weather information

19. Backing up a combination vehicle is otherwise known as ________.
A. Reversing
B. Automatic braking
C. Speeding up
D. Careful driving

20. Poor backing is responsible for ________ of road accidents involving combination vehicles.
A. 20%
B. 40%
C. 35%
D. 25%

21. The most common areas where backing-related accidents occur are ________.
A. Residential driveways and highways
B. Highways and parking lots
C. Parking lots and residential highways
D. Residential driveways and parking lots

22. What is the GOAL approach?
A. A definite place to back up your vehicle
B. Inspecting your environment while backing up
C. An approach that highlights the importance of using seat belts
D. Paying attention to potential dangers during driving

23. Who is a spotter?
A. Someone who spots law enforcement agencies in advance
B. Someone who checks your environment for potential obstacles while backing up
C. Someone who helps with identifying problems in a combination vehicle
D. Someone who helps you identify the best place to park your vehicle

24. It is advisable to tap your horn ________ while reversing.
A. Twice
B. Four times
C. Eight times
D. As many times as necessary

25. To prevent reversing-related accidents and problems, install a ________ in your vehicle.
A. Distance alarm
B. Rear-vision camera
C. Contact sensor
D. All of the above

30 Tanker Test

1. The ... regulates the activities of tanker drivers across the United States.
A. Federal Motor Carrier Safety Administration.
B. Federal Motor Safety Administration.
C. Federal Motor Company Security Administration.
D. Federal Motor Security Administration.

2. The FMCSA is an arm of the ...
A. United States Department of Security and Health.
B. United States Department of Transportation.
C. United States Department of Road Safety.
D. United States Department of Transportation and Road Safety.

3. The FMCSA's primary mission is to ...
A. Reduce the number of tanker drivers across the country.
B. Reduce vehicle-related accidents in the United States.
C. Control the operation of tanks.
D. All of the above.

4. The FMCSA outlaws the ...
A. Transportation of a significant quantity of liquids or gases without the right training and endorsement.
B. Transportation of goods across the country in a commercial vehicle.
C. Driving of commercial vehicles above a certain age.
D. All of the above.

5. The term "tanker" covers all of the following vehicles except:
A. Box trucks.
B. Flatbeds.
C. Reefers.
D. Armored tanks.

6. One of the conditions for a CDL with a tanker endorsement is ...
A. You must only drive empty tankers.
B. The tanker must be loaded.
C. It must be a six-wheel tanker.
D. The tanker's container must have the capacity to carry over 500 gallons.

7. It is illegal to carry over ... gallons of liquid in the container.
A. 2,000.
B. 5,000.
C. 8,000.
D. 10,000.

8. One of the exceptions to the above rule is that the trailer must ...
A. Be a new trailer.
B. Be driven by a driver with over five years of driving experience.
C. Be a temporary attachment to a flatbed trailer.
D. All of the above.

9. Under what condition can you drive a tanker with a Commercial Learner's Permit?
A. The tank must be empty.
B. You must have a tank endorsement on the CLP.
C. The tank must be cleaned and purged if it has been used for transporting hazardous materials in the past.
D. All of the above.

10. Some of the factors that determine the quantity of liquid you can carry with a tanker are:
A. Load temperature and the liquid's expansion rate.
B. The liquid's expansion rate and cost of transportation.
C. Load temperature and viscosity.
D. None of the above.

11. What is GCWR?
A. Gross Container Weight Record.
B. Gross Container Weight Rating.
C. Gross Combination Weight Record.
D. Gross Combination Weight Rating.

12. Some tankers can't be fully loaded. Why?
A. They are specifically designed that way.
B. They increase in volume during transportation.
C. They are at their maximum weight limit in their half-full state.
D. All of the above.

13. The tanker regulations to effect across the United States in ...
A. March 2007.
B. March 2017.
C. March 2015.
D. March 2012.

14. The penalty for violating tanker regulations is ...
A. $500.
B. $1,000.
C. $2,000.
D. $5,000.

15. You may be suspended for ... if you are caught violating the rules for operating a tanker.
A. 15 days.
B. 30 days.
C. 10 days.
D. 90 days.

16. Rollovers are common in tankers due to their ...
A. Poor steering.
B. Bad wheels.
C. High center of gravity.
D. None of the above.

17. A common problem tanker drivers must contend with is ...
A. Bouncing heads.
B. Bulkheads.
C. Bolt heads.
D. Bald heads.

18. The movement of a tanker's contents from one side to another is called ...
A. Surging head.
B. Bulkhead.
C. Surge.
D. Bolt head.

19. Spreading the weight of a tanker's load is a solution to ...
A. Bulkhead.
B. Surge.
C. Contamination.
D. Rollover.

20. The tanker endorsement test covers a wide range of areas that include all of the following except:
A. Maintenance.
B. Safety procedures.
C. Loading and offloading.
D. Inspection.

21. Are tanker drivers permitted to carry any quantity of liquid?
A. It's at the driver's discretion.
B. No.
C. Yes.
D. It depends on several factors.

22. The FMCSA was established on ...
A. December 25, 2006.
B. January 15, 2001.
C. January 1, 2000.
D. March 10, 2002.

23. The ... emphasized the importance of road safety measures prior to the enactment of the FMCSA.
A. Motor Safety Improvement Act.
B. Motor Carrier Safety Improvement Commission.
C. Motor Carrier Safety Improvement Act.
D. Motor Carrier and Safety Improvement Commission.

24. Does the type of content you transport with the tanker count in favor of or against you during the test?
A. It sometimes counts against you.
B. No.
C. Yes.
D. It always counts in favor of you.

25. When can surge trigger rollover?
A. When driving on sand.
B. When driving on a watery road.
C. When driving on ice.
D. When driving an empty tank.

31 Passenger Transport Test

1. The minimum number of people a passenger vehicle can transport is ...
A. 20 people.
B. 15 people.
C. 16 people.
D. 25 people.

2. The P Endorsement is for operators of all of the following types of vehicles except:
A. Service buses for factories and church buses.
B. Taxis.
C. Buses.
D. All of the above.

3. Another important test you must pass is the ...
A. Safety Endorsement Test.
B. Security Endorsement Test.
C. Loading Endorsement Test.
D. Air Brakes Endorsement Test.

4. Under what conditions can you use a passenger vehicle's emergency exit?
A. During a fire.
B. After a crash.
C. During incidents that can cause injuries or loss of life.
D. All of the above.

5. When is it advisable to consider an alternative to evacuation?
A. During a crash.
B. During a collision that leads to a loss of life.
C. During a fire.
D. During incidents that don't lead to a loss of life or serious injuries.

6. The FMSCA recommends testing a passenger vehicle's emergency exit every ...
A. 50 days.
B. 30 days.
C. 60 days.
D. 90 days.

7. When is it appropriate to show passengers the emergency exit?
A. During a fire.
B. Prior to a crash.
C. At your convenience.
D. As soon as you are ready to start driving.

8. What is the first step during an emergency?
A. Evacuate the passengers without delay.
B. Conduct a row-by-row evacuation.
C. Alternate evacuating the sides of the bus.
D. Inform the passengers of the need for evacuation.

9. Is it advisable to keep the emergency exit locked if passengers are aboard?
A. Yes.
B. No.
C. It doesn't matter.
D. It's based on personal preference.

10. Under what conditions should you keep the emergency exit locked while in transit?
A. When transporting people in police custody.
B. When contracted by law enforcement agencies.
C. All of the above
D. None of the above.

11. Why is it advisable to direct the passengers to a safe place during an emergency?
A. For easy accountability.
B. For convenience.
C. For no real reason.
D. A and B.

12. A minimum of ... driving experience is required to be eligible for a passenger bus license.
A. Three years'.
B. Five years'.
C. Four years'.
D. Ten years'.

13. Which of the following will disqualify you from taking the passenger transport test?
A. Previous conviction of acts of terrorism.
B. Previous conviction of driving under the influence.
C. None of the above.
D. All of the above.

14. Why is conversing with a passenger not advisable while driving?
A. It makes the passenger more abusive.
B. You lose your self-respect.
C. You may be distracted.
D. B and C.

15. Under what conditions should you not refuel your vehicle?
A. In closed buildings.
B. When passengers are on board.
C. None of the above.
D. All of the above.

16. When should you inspect your vehicle?
A. Before a trip.
B. After a trip.
C. Both before and after a trip.
D. At your convenience.

17. ... drivers are required to write up a formal inspection report after each trip.
A. Interstate drivers.
B. Intrastate drivers.
C. Independent drivers.
D. New drivers.

18. How does the brake-door interlock work?
A. It is remotely controlled by the driver.
B. It automatically engages when the door opens.
C. It opens and closes at will.
D. It allows the driver to stop the vehicle without using the traditional brakes.

19. The brake-door interlock complements the efforts of the ...
A. Parking brake system.
B. Emergency brake system.
C. Traditional brake system.
D. All of the above.

20. It is unwise to ... when using the brake-lock interlock.
A. Apply the parking brake.
B. Apply the emergency brake.
C. Take your foot off the brake pedal.
D. Switch off the lock remotely.

21. A safe oil level is when the gauge is between ... and ...
A. Full and add levels.
B. Add and full levels.
C. Intermediate and full levels.
D. Intermediate and add levels.

22. A functional power steering system should not be ...
A. Leaking when parked.
B. Leaking when in use.
C. Leaking when tightly screwed.
D. A and C.

23. A functional alternator has a tension level between ... and ...
A. ½ inch and ¾ inch.
B. 2/3 inch and 4/5 inch.
C. ½ inch and 2/5 inch.
D. 2/5 inch and 3/5 inch.

24. A good air compressor must:
A. Have sufficient air pressure and the belt mustn't be frayed.
B. Be well lubricated and have enough air pressure.
C. Not be frayed or well lubricated.
D. Have sufficient air pressure and a frayed belt.

25. To qualify for the P endorsement, an applicant must be at least ...
A. 18 years old.
B. 21 years old.
C. 30 years old.
D. 35 years old.

32 School Bus Test

1. Before you are considered eligible to drive a school bus, you must pass the ...
A. School Bus Management Test.
B. School Bus Driving Test.
C. School Bus Endorsement Test.
D. School Bus Driving and Endorsement Test.

2. Before you become a school bus driver, you must ...
A. Receive practical instruction pertaining to school bus driving.
B. Have customer service skills.
C. Understand the rules and regulations guiding school bus driving.
D. All of the above.

3. Which of the following medical reports are you required to present in order to be eligible to drive a school bus?
A. A Medical Examiner's Certificate.
B. A Comprehensive Medical Examiner's Certificate.
C. A Medical Examination Certificate.
D. A Comprehensive Medical Examination Certificate.

4. How often must you submit the required certificate?
A. Every five years.
B. Every three years.
C. Annually.
D. Every two years.

5. What is the significance of the medical test?
A. To teach you how to handle medical emergencies while driving.
B. To ensure you are physically and mentally competent to handle the job.
C. To prepare you to communicate with customers.
D. To improve your marketability.

6. When are students more prone to accidents in school buses?
A. When taking a selfie inside the bus.
B. During the loading stage.
C. During the unloading stage.
D. During both the loading and the unloading stages.

7. When are students most at risk?
A. While inside school buses.
B. While getting on or off school buses.
C. While walking towards school buses.
D. While walking away from school buses.

8. The danger zone refers to the ...
A. Areas where students are picked up or dropped off from the school bus.
B. The areas where students are taken during emergencies.
C. The areas around the vehicle where students are more prone to accidents.
D. None of the above.

9. What effect does the danger zone have on drivers?
A. It makes them sleepy.
B. The driver's vision may be blocked.
C. It makes them more forgetful.
D. It impairs their sense of hearing and speech.

10. Define the blind spot.
A. This is somewhere around the school bus where accident rates are high.
B. It refers to 10 feet on either side of the vehicle where the driver is obstructed.
C. It refers to areas behind the school bus with obstructed visibility.
D. All of the above.

11. The regulatory body that recommends special training for drivers is the ...
A. National Council for Drivers' Awareness.
B. Department of School Transportation.
C. National Safety Council.
D. National Department of School Transportation and Safety.

12. The factors that must be considered when considering the best training program in a state include ...
A. The federal government's stance on school bus driving.
B. Each school's needs.
C. Each driver's individual circumstances.
D. All of the above.

13. A safe distance for activating the warning flasher system is ...
A. 50 feet before the stop zone.
B. 10 feet before the stop zone.
C. 200 feet before the stop zone.
D. 100 feet before the stop zone.

14. Picking up and dropping off should be done at ...
A. Convenient spots for the students.
B. Convenient spots for the driver.
C. Designated drop-off and pick-up points.
D. A and C.

15. The safest place for students to cross the road after drop-off is ...
A. In front of the school bus.
B. Behind the school bus.
C. Anywhere around the school bus.
D. Wherever the students choose.

16. Check all of the following places except ... when leaving the drop-off point.
A. All side mirrors.
B. Crossover for approaching students.
C. Crossover for exiting students.
D. None of the above.

17. Why must you take a walk through the bus after completing a route?
A. You don't need to.
B. To check for sleeping students and forgotten articles.
C. To check for acts of vandalism.
D. Both B and C.

18. What steps must you take before you open the school bus door for loading or unloading?
A. Put the vehicle in neutral and engage the parking brake.
B. Put the vehicle in neutral and engage the emergency brake.
C. Put the vehicle in neutral and engage the interlocks.
D. Put the vehicle in neutral and engage the emergency blinkers.

19. For additional safety, what should you do before you move your vehicle after unloading or loading?
A. Silence the passengers.
B. Look for signs of approaching students or vehicles.
C. Turn off your radio and other noisy equipment.
D. Both A and C.

20. Which of the following items can cause an accident in the bus' doorway?
A. A book bag.
B. Jacket straps.
C. A backpack.
D. All of the above.

21. When is it advisable to give students some practical safety tips?
A. At the beginning of the school season only.
B. Towards the end of the school season.
C. At the beginning and throughout the school season.
D. At the beginning and the middle of the school season.

22. Picking students up at corners shortly before making a right turn is ...
A. A safe driving habit.
B. Something that exposes you and your passengers to risks.
C. A skill that should be practiced until mastered.
D. A criminal offense.

23. What should you do after each trip?
A. Park the vehicle on the school's premises and go home.
B. Conduct a post-trip inspection.
C. Close the windows and doors and park the vehicle in the school's garage.
D. Do whatever is convenient for you.

24. During a post-trip inspection, inspect all of the following bus parts except:
A. The stop signal arms.
B. The warning lights.
C. The mirror systems.
D. None of the above.

25. When you notice something out of place during an inspection, what should you do?
A. Notify law enforcement immediately.
B. Notify your supervisor or the school immediately.
C. Fix the problems before parking the vehicle.
D. Record your observation and keep the record for future reference.

25 Answers for General Knowledge Test

(1) (D) Two or more vehicles with a combined weight of over 26,000 pounds.

The Class A license allows its holder to drive two or more vehicles. The combined weight of the two vehicles should exceed 26,000 pounds.

(2) (C) SUVs.

As a Class A driver's license holder, you can operate a wide range of commercial vehicles. These include tractor-trailers, flatbeds, tanks, school bus and passenger buses with a specific number of passengers. This license class doesn't cover small vehicles, family cars and SUVs.

(3) (C) 16 passengers.

The P Endorsement allows holders to drive a passenger vehicle that can transport a minimum of 16 passengers. You must pass the Passenger Transport Road Skills Test and the Passenger Transport Knowledge Test in order to receive the endorsement.

(4) (C) The Transportation Safety Administration.

TSA is the acronym for the Transportation Safety Administration. The administration is responsible for the safety of the roads, drivers and pedestrians, among others.

(5) (A) Tanks.

There are different types of endorsements. They include the N Endorsement, P Endorsement and a host of others, all allowing holders to operate various vehicles.

(6) (C) S Endorsement.

The S Endorsement is for individuals who want to drive school buses. Before you get the endorsement, you must pass the relevant tests.

(7) (D) New Hampshire.

Nearly all states allow drivers to operate double or triple trailers provided they have the required endorsement. However, New Hampshire is the lone state that doesn't allow the operation of triple vehicles.

(8) (B) Class B.

The Class B license is issued to drivers who operate commercial vehicles with a combination weight of over 26,000 pounds. Licensed drivers can drive box trucks, straight trucks, tractor-trailers and segmented buses.

(9) (D) Class C.

If you have the appropriate endorsement on your Class B driver's license, you can drive vehicles that are usually for Class C license holders. This is because endorsements offer the holders some exceptions that allow them to operate vehicles under other license classes. Check the available endorsement for your license.

(10) (B) Class B.

You can drive trucks, segmented buses, large buses and straight trucks with the Class B license. Other vehicles you can operate with this class of license include delivery trucks, tow trucks and garbage trucks. You can also tow a vehicle that weighs less than 10,000 pounds.

(11) (A) Class C.

If you wish to transport hazardous materials, you can apply for the Class C license. With the license, you can transport flammable liquids, a wide range of gases, combustible liquids and explosives. You can also transport propane, gasoline, chlorine, hydrochloric acid, ammonia and sulfuric acid.

(12) (B) Eight sections.

The Commercial Driver's License Test is divided into eight sections. These are General Knowledge, Air Brakes, Hazmat, Combination Vehicles, Tankers, Doubles and Triples, Passenger Transport and School Buses.

(13) (D) Private cars.

Although there are different sections of the Commercial Driver's License test, it doesn't cover private cars.

(14) (B) 80%.

A driver is deemed qualified to receive the Commercial Driver's License if he/she scores a minimum of 80% on the exam. Anything short of that is considered an automatic disqualification from obtaining the license. This underscores the importance of thoroughly preparing for the test.

(15) (B) A candidate who gets 45 answers right.

Only this candidate passed the exam. The rule stipulates that a candidate must score at least 80%. If a candidate gets 45 answers right out of 50, that's 90%.

(16) (C) 60 minutes.

You have 60 minutes to take the test.

(17) (A) Proof of identity and proof of social security number.

Proof of identity and proof of social security number are two of the acceptable documents you should take with you to the testing centers.

(18) (A) Credit history.

There is a long list of documents that you can present as proof of identity. This includes a government-issued identity card, your birth certificate and U. S. military discharge papers. Your credit history is not considered proof of identity.

(19) (D) Duplicate documents and original copies of your documents.

Photocopies of your documents are not acceptable. Duplicates and original documents are acceptable.

(20) (A) Certificate of citizenship and birth certificate.

You need your birth certificate and certificate of citizenship to prove your permanent residency or citizenship. You can't prove your citizenship or residency with other documents such as a work permit or utility bills. Contact your local DMV for a list of acceptable documents.

(21) (A) IRS W-2 form.

The IRS W-2 form contains a list of other documents you can use to prove your residency. The form is otherwise known as an income tax return.

(22) (C) The DMV must decide.

When filling out the Commercial Driver's License application online, make sure that your name matches the one on your ID or any other document you intend to use as proof of identity. If the names don't match, you may be disqualified from taking the test.

(23) (B) Two weeks.

If you hold a Commercial Learner's Permit and wish to apply for the Commercial Driver's License, you must wait for 14 days.

(24) (B) 21 years.

In most states, you must be at least 21 years old before you are considered eligible for a driver's license. Such states may issue you a learner's permit if you are between 18 and 20 years old. However, your driving will be restricted to your state of residence.

(25) (C) 18 and 20

Applicants between ages 18 and 20 are restricted to intrastate transports until they are 21. At that age, their learner's permit will be replaced with a driver's license to enable them to operate across the country.

(26) (C) 21 years.

If you hold a Commercial Learner's Permit that restricts your activities to your state of residence, it expires once you are 21 years old. At the age, you have become an adult and can drive commercial vehicles across the country with your driver's license.

(27) (C) 21 years.

The minimum age for interstate driving is 21. Before 21, you won't be allowed to drive outside your state with your learner's permit.

(28) (D) None of the above.

The examination body can disqualify applicants based on several factors. If you can't produce proof of identity on request, your application will be disqualified. The same applies if you can't speak or write English fluently or if you present an expired ID.

(29) (C) The driver can't overturn the decision.

If you have been disqualified from driving in the past for any serious offense, you are automatically banned for life. You can't overturn the decision, meaning that you will never be allowed to get behind a wheel again.

(30) (A) 20/40.

Commercial drivers must have 20/40 vision in each eye. This is to prevent vision-related accidents.

(31) (B) 140 degrees or higher.

For horizontal vision, a driver is expected to have 140 degrees or higher. The goal is to ensure that such a driver won't have issues with seeing horizontal objects.

(32) (D) All of the above.

You may be disqualified from taking the exam if you are found with electronic devices such as cell phones or notes. Entering the test center with personal effects is also an invitation to disqualification.

(33) (A) A year-long ban from applying for the license.

A year-long ban awaits drivers that are guilty of driving under the influence of drugs or alcohol. This is to serve as a deterrent to other drivers who may be considering driving under the influence. It also ensures the security of drivers, passengers, pedestrians and other motorists.

(34) (C) Below 0.04.

The acceptable blood alcohol content for drivers of commercial motor vehicles is below 0.04. A blood alcohol content that exceeds this limit may result in stiff penalties.

(35) (D) 60 days to life ban.

The offender may be banned from driving for 60 days. However, depending on various factors, he/she may also face a lifetime driving ban.

(36) (D) You will be banned for a year.

The government penalizes drivers who use their vehicles for illegal or criminal activities. Such a crime is a felony. If you are found guilty of committing this crime with a commercial vehicle, you stand the risk of being banned from driving a commercial vehicle for a year.

(37) (A) A lifetime ban.

If you are convicted of manufacturing or distributing drugs illegally with a commercial motor vehicle, the penalty is a lifetime driving ban.

(38) (D) A year-long ban.

If you are involved in an accident that leads to a loss of life or injury, you must stop at the scene of the accident. Fleeing the scene may result in a one-year driving ban.

(39) (A) Double or triple trailers.

The endorsement allows you to drive either a double or triple trailer. As an applicant for the T endorsement, you must pass the Doubles/Triples Knowledge Test.

(40) (C) Slowly release the accelerator.

The first action when your truck starts to slide on ice should be to slowly release the accelerator. This will help reduce the vehicle's momentum and allow the tires to regain traction with the road surface, which is crucial for maintaining control. Applying the parking brake, turning the steering wheel in the opposite direction of the slide, or swiftly changing gear to reverse can make the situation worse by destabilizing the vehicle further.

(41) (A) It can cause the truck to slide further.

Using the parking brake when a truck is sliding on ice should be avoided because it can cause the truck to slide further. The parking brake, which operates on the rear wheels, can lock them up, reducing their grip and leading to a longer and potentially uncontrollable slide. The other options do not typically result from applying the parking brake in a slide.

(42) (B) When the truck is at a complete stop.

It is considered safe to use the parking brake on a truck sliding on ice when the truck is at a complete stop. Applying the parking brake while the truck is moving, especially in icy conditions, can cause the vehicle to lose control. When the truck is stopped, the parking brake can be used to prevent the vehicle from moving unexpectedly.

(43) (B) Steer in the direction of the skid and lightly pump your service brakes.

To regain control of your truck if it begins to slide on ice, you should steer in the direction of the skid and lightly pump your service brakes. This technique helps to balance the vehicle and allows the tires to regain traction. The other options - applying the parking brake quickly, increasing speed, or making a sharp turn - could exacerbate the slide or lead to a loss of control.

(44) (B) The vehicle may start to jackknife.

Applying the parking brake abruptly while your truck is sliding on ice can cause the vehicle to start jackknifing. This is a dangerous situation where the trailer swings out to form a V shape with the tractor, which can lead to a loss of control and potential accidents. The other options are not typical results of abruptly applying the parking brake on ice.

(45) (C) A Commercial Learner's Permit.
CLP means Commercial Learner's Permit. This permit is usually issued to applicants between 18 and 20 and below the 21-year age requirement for the Commercial Driver's License. With this permit, a driver can operate commercial vehicles within the state until the driver is 21 years old.

(46) (C) Maintain your speed and steering control.
In the event of a blowout while driving your truck, the first action you should take is to maintain your speed and steering control. This helps you keep the vehicle stable and prevent it from veering off course. Steering hard to the right, braking immediately, or accelerating could lead to a loss of control, making the situation more dangerous.

(47) (B) There is an unusual vibration or thumping noise.
A common sign of a truck tire going flat while driving is the presence of an unusual vibration or thumping noise. This noise is often the result of the flat tire's tread slapping against the road. The other options, such as the truck accelerating faster, reduced fuel consumption, or uncommanded steering, are not typical symptoms of a flat tire.

(48) (D) All of the above.
In the case of a flat tire, it is important to inspect the other tires on the same axle to check if they are also flat, if they need to be replaced, and to see if they have been affected by the heat or debris from the blown tire. When one tire goes flat, the others on the same axle often bear more weight, which could lead to additional tire failures if not addressed.

(49) (A) To ensure it is inflated to the manufacturer's recommended pressure.
After changing a flat tire on your truck, you should immediately check the tire pressure of the new tire to ensure it is inflated to the manufacturer's recommended pressure. Proper tire inflation is crucial for safe vehicle operation, fuel efficiency, and tire longevity. The other options are not the primary reasons for checking tire pressure immediately after a change.

(50) (A) Avoid braking until the truck has slowed down.
If your truck's tire blows out while you're driving, you should avoid braking until the truck has slowed down. Applying brakes immediately after a tire blowout can cause the vehicle to skid or lose control. Once the truck has slowed down, braking can be applied gently. You should still change lanes cautiously once control is regained, use hazard lights to signal an issue, and steer in the direction of the pull to control the vehicle.

26 Answers for Air Brakes Test

(1) (A) Compressed air brake system.

Another name for air brakes is a compressed air brake system. This is classified as a friction brake and is commonly found in vehicles that use compressed air for braking.

(2) (A) Compressed air.

While some brake systems are powered by hydraulic fluid, the air brake system is powered by compressed air. The compressed air presses on the piston and applies the brake to the brake pad.

(3) (A) A brake pad.

The brake pad is the part of the brake system responsible for stopping the vehicle. When pressure is applied on the brake pad, it exerts pressure on the brakes and the brakes are applied. When there is no pressure on the brake pad, the brakes are released.

(4) (A) Air brakes are not affected by leaks while hydraulic brakes are.

The primary difference between vehicles that use brakes powered by hydraulic fluid and those with brakes powered by air is that while hydraulic fluids may leak and lead to brake failure, air brakes are not affected by leaks. Hence, brake failure is limited to the barest minimum.

(5) (B) The air brake system.

The air brake system is noted for its reliability. Drivers can rest assured that they won't run out of fluid to power their brakes. This makes this type of brake more reliable than others.

(6) (C) George Westinghouse.

Air brakes were initially developed for use in railways by George Westinghouse. Thanks to their reliability, air brakes gradually became an integral part of modern commercial motor vehicles.

(7) (C) March 5, 1872.

After inventing the air brake system, George Westinghouse patented it on March 5, 1872.

(8) (B) Foundation brakes.

Air brakes are otherwise known as foundation brakes. They are found in buses and trucks. This braking system adopts the triple-valve principle that is common in rail cars. It can cause the exhaustion of the brake cylinder and thus release the brakes in the process.

(9) (B) Air reservoir tanks.

The air reservoir tanks are the components of the air brakes that hold enough air for the brakes to use repeatedly. As long as there is air in the air reservoir tanks, the brakes will function correctly.

(10) (C) The air compressor.

The air compressor is the air brake component that pumps the air the brake needs for operation into the air reservoir tank.

(11) (A) It connects the push rod and the brake through its arm.

The slack adjuster plays a crucial role in the operation of the air brakes. It connects the push rod and the brake through its arm. This makes it possible to control the distance between the two brake shoes. With the reduced distance between them, braking is easier and faster.

(12) (A) The foot valve.

Another name for the brake pedal is foot valve. It releases air from the reservoir tanks. When you remove the pressure by lifting your foot, the brake is disengaged and the air pressure is reduced. When you release the brakes, the system releases some air.

(13) (B) Alcohol evaporator.

An alcohol evaporator is the air brake system component that is used for putting alcohol into the system. This comes in handy in colder climates. It reduces the risk of the air brake valves freezing over.

(14) (A) Water and oil.

Water and oil are two major contaminants common in air tank drains. They may settle at the bottom of the tank and cause a malfunction in the braking system. Thus, it is imperative that you drain the air tank regularly.

(15) (B) Manual air tanks.

A challenge that is common to air tanks during cold weather is the freezing of water in the tanks. This can lead to brake failure. Manual air tanks come fully equipped with electric heating devices to prevent freezing.

(16) (A) They provide the mechanical force that prevents accidental leakage of air pressure.

Truck tractors, trucks and buses are all equipped with parking and emergency brakes. They are held together by mechanical force to prevent accidental leakage of the air pressure. Spring brakes provide the air brake with the mechanical force it needs to prevent accidental leakage of the air pressure.

(17) (C) Between 20 and 30 psi.

Straight trucks and tractor have spring brakes too. The brakes are fully operational when the air pressure in the brake drops to between 20 and 30 psi. At that stage, it may not be advisable to wait until the brakes are automatically applied.

(18) (A) Stop the vehicle while you can still control the brake.

As soon as the buzzer and low air pressure warning light flash, you should stop the vehicle. Waiting until a more convenient time could lead to an accident.

(19) (A) How the brakes are adjusted.

The braking power is dependent on how the brakes are adjusted. In badly adjusted brakes, the regular brakes won't work. The parking/emergency brakes will malfunction as well.

(20) (C) Automatic valve.

There are two types of drain valves. One is operated manually by pulling a cable or turning a quarter turn. After completing your activities for the day, drain the tanks. The automatic valve expels oil and water automatically.

(21) (B) The water in the tank may freeze.

The major cause of brake failure in cold weather is water freezing in the air tank. This problem has been corrected with the introduction of automatic air tanks equipped with electric heating devices.

(22) (A) Brake S-Cam.

The brake, shaped in the form of the letter "S," separates the brakes by pushing them apart and pushing them against the brake drum. Approximately 85% of vehicles that use air brakes in the US are equipped with the S-Cam.

(23) (C) The push rod.

The push rod is a steel rod in the air brake that shares some similarities with the piston connecting the brake chamber and the slack adjuster. When the rod is depressed, the brakes are immediately released. Extending the push rod automatically applies the brakes.

(24) (C) The brake shoe.

The brake shoe is a lined steel mechanism that produces friction when it comes in contact with the brake drum. It uses friction to stop a moving vehicle when the brakes are applied. Thus, it is crucial to the efficiency of the braking system.

(25) (A) It prevents brake malfunctioning by preventing contaminants from affecting the brakes.

Brake failure can be caused by two factors. Contaminants such as water and oil may affect the efficiency of the braking system when they settle at the bottom of the air tank. By draining the air tank regularly, such contaminants can't settle.

27 Answers for Doubles and Triples Test

(1) (A) These are long vehicles with two or more trailers attached to the truck's tractor.

Although there are different types of combination vehicles, the doubles and triples stand out. In the former, two trailers are attached to the truck's tractor while three or more trailers are attached to the tractor in the latter.

(2) (C) Large Commercial Vehicles.

Double and triple trailers are otherwise known as Longer Commercial Vehicles. It is apparent from their names that this set of vehicles is longer, bigger and weighs more than other vehicles in the commercial vehicle class. Thus, this class of vehicle requires extra driving skills to operate.

(3) (B) No.

States don't have a consensus agreement on restrictions on triples. Each state defines its own rules that govern combination vehicles, triples inclusive. Thus, it is imperative that you are familiar with the rules and regulations governing the activities of commercial vehicle operators in your state of residence.

(4) (B) It is a test of your ability to fix whatever you uncouple.

Sometimes, you may need to uncouple your vehicle, either to detect a problem or fix one. If you can't re-couple the vehicle, you run the risk of operating a badly coupled vehicle, which is a time bomb.

(5) (C) Handling.

Handling is a test of your ability to operate a vehicle irrespective of the potential challenges. Since driving is not always without occasional challenges, your ability to handle these problems will enhance your driving skills.

(6) (C) None of the above.

When applying for the test, you must pay a testing fee and licensing fee. Your state may also require some additional fees.

(7) (A) Your state of residence.

Since drivers take the test in their location of choice and states have different requirements, you have a better chance of meeting the requirements of your state of residence than somewhere you are not familiar with.

(8) (C) They are less stable than other vehicles.

When driving doubles or triples, you have to consider their instability. They are notably less stable than other vehicles, one of the reasons why they are prone to rolling over, jackknifing and other related problems.

(9) (D) Both B and C. Driving slowly and steering gently.

Rolling over is common to long vehicles. It occurs when the vehicle loses balance and rolls over, perhaps due to speeding or abrupt steering control. You can prevent this by driving slowly and steering gently.

(10) (B) When a trailer driver changes lanes quickly.

The crack-the-whip effect usually occurs when a trailer driver changes lane quickly. The tractor and the attached trailers will feel the effect of the abrupt lane change differently. While a section of the vehicle may not feel the impact too much, another part may feel it, leading to rollover.

(11) (D) To make it easier for the driver to maneuver the vehicle.

Long vehicles are usually more difficult to maneuver than their shorter counterparts. Hence, you must consider this factor when driving. Don't follow a vehicle so closely that you find it challenging to avoid an accident.

(12) (A) Skidding and loss of traction.

Two problems that are common to doubles and triples are skidding and loss of traction. They are responsible for most of the accidents involving this type of vehicle. Sudden brake application can cause skidding, which invariably causes jackknifing too.

(13) (A) Ease of driving out of the parking lot.

When parking a long vehicle, consider the ease of driving out of the parking lot. Unlike with smaller vehicles, you need plenty of space.

(14) (A) Securing the trailer.

An important step towards ensuring your trailer's emergency brakes function properly is securing the rear trailer.

(15) (B) Behind the tractor.

The semitrailer with the heaviest load should be connected behind the tractor. In that position, it is easier to pull the trailer. More so, it offers the combination vehicle the needed stability that will prevent it from jackknifing or rolling over when in motion.

(16) (D) In front of the rear trailer.

The best place to put the converter dolly is in front of the rear trailer. In this position, it is able to connect the tractor and trailer with relative ease. This contributes significantly to the overall stability of the vehicle.

(17) (A) By locking the pintle hood and securing the converter gear support.

Place the first semitrailer in front of the converter dolly and hook it to the front trailer by locking the pintle hook and securing the converter gear support while in a raised position.

(18) (A) Lower and upper fifth wheel.

While coupling your trailer, ensure that there are no spaces between the lower and upper fifth wheel. The locking jaws should also be closed on the kingpin before connecting the air hoses, safety chains and light cords. Once you are done with that, close the air tank petcock.

(19) (C) By pushing the air supply knob in.

You can charge the brakes by pushing the air supply knob in. Once you are done with that, you can check the second trailer for air by opening the emergency line shutoff. If there is no air pressure, the brakes won't work.

(20) (A) It makes the vehicle safer to drive.

There are several benefits to carrying out routine inspections on your vehicle. In general, such regular inspections help you to identify and fix potential problems before they get out of hand.

(21) (C) All of the above.

You can take the test either at the local DMV or a third-party site. While some states are comfortable with third-party sites, others make it mandatory that the tests are done via the department. Hence, your state of residence will play a huge role in your choice.

(22) (D) Longer Combination Vehicles.

LCV stands for Longer Combination Vehicles. It is a collection of vehicles that are not only bigger than other vehicles, but also longer and heavier as well. In this group are tankers, heavy school buses, large vehicles for transporting hazardous materials and a host of others.

(23) (D) Purchasing and supply.

When taking the doubles and triples test, your examiner will test your knowledge of handling, assembly procedures and potential problems. Note, however, that it is at each state's discretion to choose the format and style of the exam. While the general concept remains the same, the content may differ.

(24) (A) The cost of purchase.

While double or triple trailers are undoubtedly expensive, that doesn't impact a driver's ability to operate them. However, their sheer size, length and outstanding weight can make them a handful for most drivers. These factors also underscore the need for competent drivers to handle them, hence the need for this test.

(25) (D) All of the above.

These are some safety precautions you should take into consideration when uncoupling the rear trailer. Removing some of its load makes it lighter and easier to uncouple. Parking the rig on level and firm ground prevents the vehicle from accidentally rolling forward or backward, which is the same reason for parking it in a straight line.

28 Answers for Hazmat Test

(1) (B) Hazmat means hazardous materials.

Hazmat means hazardous materials. This is in reference to substances that are injurious to health and can damage property and/or be a serious threat to public health.

(2) (A) Matches and fuses.

Matches and fuses are some typical examples of flammable solids that can constitute serious health hazards. The list of hazardous materials includes flammable liquids, a wide range of gas, combustible liquids, explosives, propane, gasoline, chlorine, hydrochloric acid and ammonia.

(3) (A) Characteristics and shipper's decision.

Characteristics and a shipper's decision are two major factors that determine whether a substance is hazardous or not. When shipping such materials, the shipper makes transporters aware of the nature of a substance by indicating whether it is hazardous or not.

(4) (B) English only.

The hazmat test is conducted in English only. Thus, a driver must understand, read and speak English before being eligible to take the test.

(5) (C) All of the above.

One of the ways you can identify a hazardous substance is by its label(s). A substance with an RQ or X label in the right column is hazardous. The shipper will also use placards or diamond labels to notify people of a substance's hazardous nature.

(6) (A) Fireworks and pest control.

Pest control chemicals and fireworks usually release hazardous materials. Hence, if you are hired to help with the distribution of these products, you will need an endorsement for transporting hazardous materials.

(7) (A) An H or X endorsement.

With either of these endorsements on your CDL, you can legally transport hazardous materials.

(8) (A) You must take part in a security threat assessment and be aged 21 and above.

You must meet some requirements to qualify for an endorsement to transport hazardous materials in the United States. Some of these requirements include passing a security threat assessment conducted by the TSA as well as being at least 21 years old.

(9) (C) 10 years.

For your CDL renewal, you are required to provide a history of your previous licenses. If possible, you must provide licenses that cover the past 10 years if you are not new in the trucking industry. This is irrespective of your state of residence or type of license.

(10) (C) A TSA Security Threat Assessment.

To qualify for the type of CDL that permits you to transport hazardous materials, it is mandatory that you pass the TSA Security Threat Assessment.

(11) (C) May 5, 2003.

The TSA test rule was made public on May 5, 2003, when the TSA published the 49 CFR 1572 rule under which the TSA is required to conduct the threat assessment. The goal is to ensure that hazardous substances are properly and safely transported across the country.

(12) (C) 30 days.

You are expected to complete the assessment within 30 days of submitting your fingerprints to the examiner. Failure to do this will result in your application being disqualified. However, the submission deadline may be extended depending on certain circumstances.

(13) (A) You have eligibility issues.

The assessment completion can be extended under two conditions. These are if you have eligibility issues or your paperwork is incomplete. In either case, the extension is to allow you or the TSA to fix the issues without delay so you can take the exam.

(14) (D) You have a criminal record.

You will be permanently barred from getting the hazardous materials endorsement if you have a criminal record. These include a record of terrorism-related crimes and others that include rape, money laundering and kidnapping.

(15) (A) Espionage and sedition.

Sedition and espionage rank very high among the crimes that may incur an automatic disqualification from getting a hazmat endorsement on your driver's license. These and other related crimes are considered serious enough to undermine the country's security and thus the examiner won't give such individuals access to harmful substances.

(16) (A) A fugitive.

As a fugitive, you can't take the test. The same rule applies to non-US citizens, individuals who are not permanent residents, those who have been medically certified as not mentally competent to handle commercial vehicles or who have involuntarily received medical attention at a mental institution.

(17) (D) None of the above.

Any applicant who has committed any of the crimes listed here can't take the test. This includes kidnappers, arsonists and hostage takers. Others who are not allowed to take the test are applicants with a tax evasion history and individuals who are guilty of damaging the environment or disrupting the transportation system.

(18) (B) Mistaken identity and reversed conviction.

You can challenge a ban if you can prove a case of mistaken identity when you were accused of a crime that got you disqualified from taking the test. You should also be able to prove that a conviction was reversed.

(19) (A) Acknowledgment of conviction based on reason of insanity.

Some extenuating circumstances allow you to apply for a waiver if you are disqualified. These are acknowledgment of conviction based on reason of insanity or acknowledgement of being involuntarily committed to a mental institution.

(20) (D) 30 days of the receipt of your waiver application.

Once you submit a waiver request, the TSA will respond within 30 days of the receipt of your application. In the meantime, you can hone your skills and keep preparing for the test.

(21) (D) Both A and B. Roadside infractions and placarding violations.

Although you may be disqualified on multiple grounds, you can't be disqualified for placarding violations or roadside infractions.

(22) (D) The Transportation Security Administration.

The Security Threat Assessment is conducted by the Transportation Security Administration. This body is responsible for regulating the activities of the transportation industry with a view to ensuring safer roads by providing guidelines and enforcing training. As drivers are trained to be more security conscious, the roads become safer.

(23) (C) State and federal governments.

To be eligible for any type of endorsement, including one that allows you to transport hazardous materials, you must meet both state and federal government criteria.

(24) (D) All of the listed signs.

To determine whether a tank is used for transporting hazardous materials or not, check the placard and diamond labels.

(25) (C) In heavy commercial vehicles.

Although there are several options when considering how to transport hazardous materials, the best and safest way to transport them is in heavy commercial vehicles.

29 Answers for Combination Vehicle Test

(1) (B). A combination vehicle test.

If you can handle a wide range of heavy commercial vehicles, you must pass the combination vehicle test. This test is specifically designed for commercial vehicle drivers who specialize in operating combination vehicles of any type.

(2) (A). They are heavier and longer.

Some attributes of combination vehicles such as longer size and heavier weight distinguish them from other vehicles in the combination class. Their sheer length and weight requires drivers to have some special driving skills before they can operate such vehicles without issues.

(3) (B). Have excellent driving experience and top-notch knowledge of the vehicle.

Even if you have operated other types of commercial vehicles, combination vehicles are a different ball game. You need special driving experience before you sit behind the wheel.

(4) (B). A written test.

While both the written and driving tests are very important, each applicant is expected to take the written test first. It is a test of your theoretical knowledge of combination vehicles before you attempt the practical test.

(5) (A). Semitruck.

A combination of vehicles is otherwise known as a semitruck or semi. It is a combination of a tractor unit and a trailer or many trailers. The connection between the tractor and the trailer(s) is made with a converter dolly or fifth-wheel dolly. Sometimes, both dollies can be used.

(6) (A). A fifth-wheel dolly.

Combination vehicles can easily bend in the middle because of the fifth-wheel dolly used for connecting the tractor unit and the attached trailer or trailers.

(7) (B). A fully-loaded truck.

While combination vehicles are more difficult to stop than a regular car, a fully loaded truck is still easier to stop than an empty truck. Thus, when handling an empty truck, you must be extremely careful when you brake.

(8) (B). It is a condition in which the vehicle bends into a V shape while skidding uncontrollably.

Jackknifing is a common problem to combination vehicles, thanks to their length. It is a condition in which a combination vehicle bends into a V or L shape. This is more common when the vehicle skids uncontrollably and may cause problems including rollovers.

(9) (A). A loss of traction.

Loss of traction is the primary cause of jackknifing. Loss of traction occurs when there is low friction between the tires and the road. This may be caused by gravel, sand, ice, snow or any other substance that causes tires to lose their grip on the road surface.

(10) (D). Both A and B. Empty trailers and fully-loaded trailers.
Jackknifing is common to empty trailers and trailers with badly distributed loads. This is because stopping an empty vehicle is usually more difficult than stopping a fully loaded one. More so, uneven distribution of the loads carried by a combination vehicle may make it unbalanced, causing jackknifing.

(11) (A). Decelerating and braking sharply.
A preventive measure against accidents when driving on a curve is to avoid braking and decelerating sharply. Either of the two actions may result in a rollover or a more serious accident.

(12) (A). Spread your braking over a distance.
An effective preventive measure against jackknifing is spreading your braking over a reasonable distance. If you don't brake suddenly, the vehicle won't jackknife.

(13) (A).Skidding.
Jackknifing doesn't occur suddenly. When you apply the brakes suddenly, the vehicle may skid. Skidding will then trigger jackknifing. Thus, any driving-related problem that can cause skidding will also increase a vehicle's tendency to jackknife. Watch out for such problems and avoid them.

(14) (C). Pay attention to your environment.
Most braking-related problems are avoidable if you take the necessary precautions. This includes paying attention to your environment, identifying potential problems and avoiding them. This includes blind spots and potholes.

(15) (A). A situation in which both the rear and front wheels take different paths.
In combination vehicles, cheating is a common problem that may lead to skidding, and by extension, jackknifing. It is a situation that arises when both the rear and front wheels take different paths. The rear wheels won't take the same path as the front wheels.

(16) (C). Focusing on the front wheels.
Cheating can be controlled and corrected. To prevent skidding or jackknifing, focus on the front wheels. If properly controlled, the rear wheels won't go out of control, even if they follow a different path.

(17) (A). Not shifting gears while crossing railroad tracks.

You may get stuck on railroad tracks when driving a combination vehicle if you don't take necessary precautions. To avoid this, under no condition should you shift gears while crossing railroad tracks. Do that either before or after crossing the tracks.

(18) (A). It contains a number you can call for immediate assistance.

If you are stuck on the tracks, you can call for immediate assistance by locating the number on the DOT sign.

(19) (A). Reversing.

Backing up is also known as reversing. Backing up a combination vehicle is very risky and should be done with extreme caution. It is not advisable to back up on a highway with such a long and heavy vehicle. You should only reverse the vehicle when absolutely necessary.

(20) (D). 25%.

The National Safety Council reports that approximately 25% of vehicle accidents can be attributed to poor reversing. Over 500 deaths and 15,000 injuries are caused by reversing-related accidents. Hence, be cautious when backing up a combination vehicle.

(21) (D). Residential driveways and parking lots.

The US National Highway Traffic Safety Administration reports that most backing-up-related accidents occur in parking lots and residential driveways. Sadly, the driver backing up is usually blamed for such accidents. This is another reason why you should back up your vehicle with caution.

(22) (B). Inspecting your environment while backing up.

The GOAL approach simply means "get out and look." The approach underscores the significance of checking your surroundings while backing up. When you have a good idea of your environment, you can easily avoid potholes, blind spots and other potential problems.

(23) (B). Someone who checks your environment for potential obstacles while backing up.

Sometimes, you may need someone to check your surroundings when you are behind the wheel. The spotter will inform you of potential problems. With effective communication between both of you, a spotter is an asset.

(24) (A). Twice.

Before you start backing up, tap your horn twice to notify other road users or passersby and avoid preventable accidents. The horn will draw people's attention to your vehicle as well as enable them to know the direction you are coming from. This can come in handy in crowded areas.

(25) (B). Rear-vision cameras.

With the cameras, you can see your vehicle's rear easily and avoid running into obstacles. Hence, your problems with rear blind spots are solved once you have full visual control of your vehicle's rear while backing up.

30 Answers for Tanker Test

(1) (A). Federal Motor Carrier Safety Administration.
The Federal Motor Carrier Safety Administration is responsible for regulating the activities of tanker drivers across the United States.

(2) (B). United States Department of Transportation.
The FMCSA is under the supervision of the United States Department of Transportation. This department's primary mission is to reduce the number of injuries caused by large buses and trucks.

(3) (B). Reduce vehicle-related accidents in the United States.
The FMCSA's primary mission is to reduce vehicle-related accidents and fatalities caused by large buses and trucks in the United States.

(4) (A). Transportation of a significant quantity of liquids or gases without the right training and endorsement.
For reasons of safety, the FMCSA places restrictions on the transportation of liquids and gases across the country with a tanker.

(5) (D). Armored tanks.
The term "tanker" covers a wide range of vehicles. This includes box trucks, flatbeds, dry vans and reefers. The FMCSA stipulates that if you drive any of these vehicles, you must have the right endorsement on your CDL.

(6) (B). The tanker must be loaded.
There are certain conditions under which your CDL must be accompanied by a tanker endorsement. One such condition is that the tanker must be fully loaded.

(7) (D). 10,000.
A tanker driver can't carry over 10,000 gallons of liquid. This prevents accidents due to overloading and other related mistakes.

(8) (C). Be a temporary attachment to a flatbed trailer.
You don't need a special endorsement to operate a temporary attachment to a flatbed trailer, to operate a tank designed for transporting goods or an empty tank.

(9) (D). All of the above.
You can operate a tanker with a Commercial Learner's Permit under certain conditions such as if the tank is empty or if you have a tank endorsement on the permit. If the tank has previously been used to transport hazardous materials, you must clean and purge it.

(10) (A). Load temperature and the liquid's expansion rate.

Restrictions are placed on the quantity of liquid a tanker can transport at a time. Some of the factors that determine the quantity include load temperature and the liquid's expansion rate. Other factors are the maximum weight limit allowed by the government on public roads and the tank's volume capacity.

(11) (D). Gross Combination Weight Rating.

GCWR means Gross Combination Weight Rating. This refers to the maximum allowed combined mass of a combination vehicle which includes both the cargo and the passengers. For a tanker, this is the combined mass of the passengers in the tanker as well as the tank's contents.

(12) (C). They are at their maximum weight limit in their half-full state.

Some loads reach the maximum weight limit, even before the tank is filled up. In such cases, the tank can't be fully loaded. Thus, the tank will have some extra space that can cause a problem while transporting liquids.

(13) (B). March 2017.

In an attempt to regulate the trucking industry, especially the tanker subsector, the government introduced the tanker regulations in March 2017. Since then, it is mandatory for tanker drivers to meet some basic requirements before they are deemed qualified to operate a tanker. Violators are seriously penalized.

(14) (D).$5,000.

In order to enforce the rules and regulations, the regulatory body fines tanker drivers without the right authorization to operate such vehicles $5,000. This is to serve as a deterrent to others and serve as a warning to others of the FMSCA's zero tolerance of violations of its regulations.

(15) (D). 90 days.

Aside from the monetary penalty of $5,000, you may also face a long suspension time of 90 days.

(16) (C). A high center of gravity.

Rollovers are common in tankers because, unlike smaller vehicles, they have a very high center of gravity. Thus, they are highly unstable and can be affected by conditions such as driving on ice, an abrupt lane change and bad weather.

(17) (B). Bulkheads.

One of the biggest challenges a tanker driver must contend with is bulkhead-related problems. A bulkhead is a barrier or a dividing wall between the compartments in the tank. Bulkheads are used for dividing large tanks into smaller tanks. They make the effects of a surge more pronounced in tankers.

(18) (C). Surge.

The movement of a tanker's contents from one side to another is known as surge. Surge is common when a tanker is stopped. The liquid will move forward in response to the stop action and hit the front of the tank, shoving the tank forward and vice versa.

(19) (A). Bulkhead.
Since bulkhead is triggered by the existence of dividing walls in a tanker, the problem can be addressed by spreading the weight of the tanker's load evenly among the compartments. This will prevent the bulkhead from affecting the tanker adversely.

(20) (C). Loading and offloading.
The tanker endorsement test covers a wide range of areas that include maintenance, safety procedures and inspection.

(21) (B). No.
Drivers can transport gaseous or liquid containers that can carry more than 119 gallons. The total volume of the containers' content exceeds 1,000 gallons.

(22) (C). January 1, 2000.
The FMCSA was established on January 1, 2000, pursuant to the Motor Carrier Safety Improvement Act of 1999.

(23) (C). The Motor Carrier Safety Improvement Act.
The Motor Carrier Safety Improvement Act emphasized the importance of road safety measures prior to the enactment of the FMCSA.

(24) (B). No.
While several factors count towards your eligibility for the test, the type of content you carry isn't one of them. The focus is on meeting other requirements as stipulated by the regulatory body.

(25) (C). When driving on ice.
Rollover is more often than not triggered by ice. Thus, when driving on ice, you should take extra care, drive slowly and steer carefully.

31 Answers for Passenger Transport Test

(1) (C). 16 people.

A passenger vehicle should carry a minimum of 16 passengers, including the driver.

(2) (D). All of the above.

The P Endorsement is a special type of endorsement for operators who desire to operate a wide range of vehicles that include service buses for factories, church buses, general buses and taxis.

(3) (D). Air Brakes Endorsement Test.

Most large vehicles, including passenger buses, are equipped with an air brake system. Thus, in addition to passing the general passenger bus test, you must also take a special air brakes test. This is to ensure that you are familiar with the operation of the brake system.

(4) (D). All of the above.

The emergency exit is designed for use in emergency situations only. It should be used in conjunction with the regular exit.

(5) (D). Incidents that don't lead to a loss of life or serious injuries.

While evacuation may be the best solution in some cases, some exceptional cases warrant the consideration of other alternatives. For instance, there isn't a need for evacuation if an incident doesn't lead to serious injuries or a loss of life.

(6) (D). 90 days.

According to the FMSCA, each driver should test the emergency exits every 90 days. This is to identify possible issues with the exits and keep them in full working order.

(7) (D). As soon as you are ready to start driving.

It is good practice to inform the passengers of the existence of emergency exits and their locations at the start of a trip, before any emergency may present itself.

(8) (D). Inform the passengers of the need for evacuation.

When an emergency situation arises, don't panic. Reassure your passengers of their safety. If you must evacuate them, explain your decision. Inform them of the need for the evacuation and share safety precautions they should take.

(9) (B). No.

It is not advisable to keep the emergency exit locked when passengers are on board. This may make it difficult to use the exits when emergency situations arise.

(10) (C). All of the above.

Although the rule of thumb suggests the emergency exit door shouldn't be locked when passengers are on board, there are some exceptions. For instance, you should keep exits locked when transporting people in police custody or when contracted by law enforcement agencies.

(11) (A). For easy accountability.

During an emergency such as a car crash, some passengers may get lost in the chaos. It is easier to know those who are missing if you direct the other passengers to a safe place. This enables the driver or other people to search for the missing passengers.

(12) (A). Three years.

You must have a minimum of three years of experience to operate a passenger vehicle. This is for your safety, the safety of the passengers and the safety of other road users.

(13) (D). All of the above.

Even if you meet the eligibility requirements for the license, you may still be denied the license under some conditions such as if you have been convicted of acts of terrorism or of driving under the influence of drugs or alcohol.

(14) (C). You may be distracted.

Distraction is the primary reason behind this restriction. Focusing on the road may be difficult when you are engaged in a serious discussion with a passenger. Your inability to pay undivided attention to the road endangers your life, that of your passengers and other road users as well.

(15) (D). All of the above.

Don't refuel in closed buildings or when passengers are on board your vehicle to prevent them from exposure to poisonous fuel vapors.

(16) (C). Before and after a trip.

It is necessary that you inspect the vehicle regularly, both pre- and post-trip, to enable you to identify potential problems and fix them before they get out of hand.

(17) (A). Interstate drivers.

Interstate drivers are required to write up a formal post-trip inspection report after each trip and submit it to their employer. The report must include vehicle type and any observations made during the inspection such as whether there are damaged parts that need to be replaced.

(18) (B). It automatically engages when the door opens.

The brake-door interlock is an automatic braking system that becomes fully operational whenever the door opens. It keeps the vehicle stationary while passengers are alighting. This is an effective preventive measure against accidents that occur when passengers are getting off a bus.

(19) (A). Parking brake system.

The brake-door interlock complements the parking brake system since it also helps keep the stopped vehicle stationary once the vehicle's doors are opened.

(20) (C). Take the foot off the brake pedal.

Some drivers erroneously consider the brake-door interlock a replacement for the parking brake. They take their foot off the brake pedal once the interlock is engaged. This is dangerous as the interlock disengages automatically when the door closes. The interlock complements the parking brake's efforts. It does not replace it.

(21) (B). Add and full levels.

When checking the oil level, it is good to know that the oil level is safe for use when it is between the add and full level. If it is below the correct level, top up the oil before driving. Otherwise, you may be taking a huge risk with the engine.

(22) (C). Leaking when screwed tight.

A check of your power steering system is also recommended during your pre- and post- trip inspection. If the power steering is functional and is in good condition, it shouldn't be leaking when screwed tight. If it does, you must identify the issue and fix it without delay.

(23) (A). ½ inch and ¾ inch.

Pay attention to the alternator during the inspection. When it is performing at maximum capacity, its tension level should be between ½ inch and ¾ inch. Any reading besides this is an indication of any error somewhere that needs urgent attention.

(24) (A). Must have sufficient air pressure and the belt mustn't be frayed.

The air compressor, as an important passenger bus component, must have sufficient pressure to perform its duties. Otherwise, it may malfunction. Its belt is another area that deserves attention. It mustn't be frayed or broken.

(25) (B). 21 years.

Until you are 21, you won't qualify for the passenger bus endorsement. This is to ensure that only people that are legally considered adults are permitted to drive such vehicles.

32 Answers for School Bus Test

(1) (C). School Bus Endorsement Test.

Before you are considered eligible to operate a school bus, your examiner must be convinced of your ability to handle the bus safely. Hence, it is mandatory that you take the School Bus Endorsement Test and pass it before your endorsement is confirmed.

(2) (D). All of the above.

Before you can be endorsed to drive a school bus you must be trained in customer service skills, practical instructions about school bus driving and understanding the rules and regulations guiding the operation of school buses. Once you are done with the program, you can take the test.

(3) (A). A Medical Examiner's Certificate.

As proof of your physical fitness, you are required to provide a Medical Examiner's Certificate as proof of your eligibility for the license.

(4) (D). Every two years.

Apart from providing the Medical Examiner's Certificate before you are licensed, you must also resubmit it every two years. This is primarily to ensure that you still maintain the physical and mental fitness that initially qualified you for the license. If your mental or physical health deteriorates, your endorsement may be withdrawn.

(5) (B). To ensure you are physically and mentally competent to handle the job.

Driving is both physically and mentally challenging. Loading and unloading, and stopping at different drop-off and pick-up points may test your physical fitness. Thus, a potential employer may be concerned about your ability to meet the challenges and will test them accordingly.

(6) (D). During loading and unloading.

Students are more prone to accidents while being dropped off by a school bus or when boarding it. This is especially true if the aisle is obstructed or school bags or other items get caught in the doorway. Sometimes, a student may be hit while in the blind spot too.

(7) (B). While getting on or off school buses.

Students are vulnerable to accidents when boarding or alighting from school buses. There are reportedly more deaths at such times than those that occur while inside school buses.

(8) (C). The areas around the vehicle where students are more prone to accidents.

These are areas where the driver's vision is blocked so he/she may not be able to see the boarding or exiting students clearly. Such areas include blind spots ten feet on either side of the vehicle and in front of the vehicle.

(9) (B). The driver's vision may be blocked.
Drivers are more vulnerable to accidents in the danger zone. This can lead to making a wrong driving decision that may trigger a crash or other accidents.

(10) (D). All of the above.
Blind spots are areas behind the school bus where drivers have obstructed visibility. These are areas around the school bus which are more prone to accidents.

(11) (C). National Safety Council.
The National Safety Council recommends special training for school bus drivers to ensure the safety of students, drivers, pedestrians, etc.

(12) (C). The driver's individual situation.
The basic requirements for eligibility for a school bus endorsement differ from state to state. Hence, each state must consider the individual needs of potential drivers.

(13) (D). 100 feet before the stop zone.
Before stopping the vehicle, notify other drivers of your intention by activating the warning light system. It's best to activate the system 100 feet before the stop zone. This helps you to avoid dangerously abrupt stops.

(14) (C). Designated drop-off and pick-up points.
Only drop off and pick up students at designated points. This not only makes it easier to discharge your duties but is safer and less time-consuming as well.

(15) (A). In front of the school bus.
The front of the school bus is the safest place for students to cross the road after drop-off.

(16) (D). None of the above.
When checking around the vehicle for people or vehicles, check the crossover for approaching students, all the side mirrors and the crossover for exiting students. Checking everywhere is an effective way to reduce accidents that occur around school buses.

(17) (D). Both B and C.
After completing a route, don't simply switch off the engine and go home. Take a walk through the bus and check for sleeping students and forgotten articles. Check for acts of vandalism too. If you spot sleeping students, wake them up immediately. Return forgotten articles to the school.

(18) (A). Put the vehicle in neutral and engage the parking brake.
Before opening the door for loading and unloading, put the vehicle in neutral and engage the parking brake. This is important. It ensures that the vehicle remains stationary during the entire loading or unloading period. Thus, the vehicle can't roll over students or other road users.

(19) (D). Both A and C.

It is important that you are aware of your environment, especially what goes on in and around your vehicle. To enhance your hearing moving, silence the passengers. Don't forget to turn off noisy equipment that may interfere with your hearing.

(20) (D). All of the above.

Sometimes, accidents are caused by some simple things such as book bags, backpacks or jacket straps or strings being caught in the bus' doorway. Items tangled in the doorway can cause serious injuries.

(21) (C). At the beginning and throughout the school season.

The students' safety is one of your major responsibilities. At the beginning of the season, give them some practical safety tips. Throughout the school season, give them constant safety reminders.

(22) (B). Expose you and your passengers to risks.

When dropping off students, doing so at corners shortly before making a right turn is not ideal as it exposes both you and your passengers to needless risks. You have no idea what is happening around the corner. Perhaps there are incoming vehicles or other things that may expose you to risk.

(23) (B). Conduct a post-trip inspection.

After completing a trip, you should conduct a post-trip inspection. Close all open doors and windows. Check for operational or mechanical problems. Pay attention to the warning lights, mirror systems, stop signal arms, etc.

(24) (D). None of the above.

The post-trip inspection involves checking some major parts of the vehicle including warning lights, stop signal arms, etc.

(25) (B). Notify your supervisor or the school.

The inspection is to ensure that the vehicle is in good condition. If you notice problems, either notify your supervisor or the school authorities without delay. Let them rectify any issues you cannot address.

33 General Knowledge Test

1. To schedule a test, you can visit the ...
A. Department of Motor Vehicles.
B. Department of Motor Licensing.
C. Department of Motor Licensing and Registration.
D. Department of Motor Registration and Licensing.

2. You can take the road test on any day except:
A. Wednesday.
B. Saturday.
C. Friday.
D. Tuesday.

3. You should arrive at the test center ... early.
A. 30 minutes.
B. 45 minutes.
C. 10 minutes.
D. 15 minutes.

4. A Rhode Island resident can register for and take the test in ...
A. Rhode Island and New York.
B. Rhode Island and Washington.
C. Rhode Island only.
D. Any state.

5. You can get a list of test centers when you visit the ...
A. Official website of the federal Department of Motor Vehicles.
B. Official website of the local Department of Motor Vehicles.
C. Official website of the local liaison department.
D. None of the above.

6. If you fail the test on the first attempt, you can still retake the test a minimum of ... on a permit.
A. Three times.
B. Four times.
C. Two times.
D. Five times.

7. Your eligibility to apply for a permit will become invalid ...
A. After two attempts.
B. After five attempts.
C. If you are physically unfit to operate a vehicle.
D. None of the above.

8. If you are under 18 and fail the test, you must wait for ... before you can retake the examination.
A. Seven days.
B. Five days.
C. As determined by the Department of Motor Vehicles.
D. As determined by the local licensing and registration committee.

9. While waiting to retake the exam, it is advisable that you ...
A. Revisit the Department of Motor Vehicles to learn why you failed.
B. Go through your study materials again and improve your driving skills.
C. Apply to take the exam in other states.
D. All of the above.

10. Before you are allowed to reschedule your test, you must first ...
A. Cancel your first appointment.
B. Pay a rescheduling fee.
C. Wait for three weeks after the first test.
D. Wait for five weeks after the first test.

11. Can a driver apply to take two tests simultaneously?
A. No.
B. Yes.
C. The DMV must decide.
D. It depends on certain factors.

12. Under what conditions can you reschedule your test appointment?
A. Change of job and inclement weather.
B. Change of job.
C. Inclement weather and tardiness to the exam center.
D. Lack of interest in the exam and change of job.

13. You will fail the test if you do any of the following except:

A. Run a red light.

B. Cut corners driving on winding roads.

C. Drive on curbs.

D. None of the above.

14. While taking the test, a reckless driver nearly gets you involved in an accident. The best reaction is to ...

A. Maintain your cool no matter what.

B. Refuse to give the right of way.

C. Respond aggressively.

D. None of the above.

15. Before the examination, which of these will improve your chances of passing the test?

A. Adequately preparing for the test.

B. Staying clear of drugs and alcohol.

C. Keeping calm before the examination.

D. All of the above.

16. To qualify for the road skills test, you must have a valid CLD for a minimum of ... before you schedule the test.

A. 45 days.

B. 30 days.

C. 40 days.

D. 60 days.

17. To take the road test, you should visit ...

A. The nearest driving school.

B. The Commercial Driver Road Test Lot.

C. The Commercial Driver Road Test Park.

D. The nearest licensing school.

18. When taking the road skills test, you should arrive at the test center ...

A. 10 minutes early.

B. 15 minutes early.

C. 5 minutes early.

D. 30 minutes early.

19. If you are under 21 and taking the road test, you must provide your ...
A. 50-hour Certification of Eligibility for Provisional License form.
B. 40-hour Certification of Eligibility for Provisional License form.
C. 60-hour Certification of Eligibility for Provisional License form.
D. 20-hour Certification of Eligibility for Provisional License form.

20. Each person taking the road skills test must be accompanied by ...
A. A guarantor.
B. A licensed driver.
C. A driving school instructor.
D. A lawyer.

21. If you drive yourself to the test center unaccompanied by the right person, you will be ...
A. Banned from taking the test for 3 months.
B. Banned from taking the test for 6 months.
C. Banned from taking the test for 9 months.
D. Banned from taking the test for 12 months.

22. A driver is required to devote a minimum of ... to practicing driving after sunset to prepare for night driving.
A. 20 hours.
B. 15 hours.
C. 12 hours.
D. 30 hours.

23. You should practice driving in both moderate and heavy traffic for ...
A. 25 hours.
B. 15 hours.
C. 10 hours.
D. 30 hours.

24. When should you stop the vehicle and bring it to a stop for the smoothest possible braking?
A. As soon as you reach the stopping position.
B. Shortly before reaching the stopping position.
C. When it is convenient for you to do so.
D. It depends on the type of vehicle you drive.

25. While shifting the gear, ... is an error you should avoid at all costs.
A. Auto shifting.
B. Grinding.
C. Frictionless grinding.
D. Abrupt shifting.

26. When approaching an intersection and asked to stop, where is the best position to stop the vehicle?
A. Before the intersection.
B. At the intersection.
C. After the intersection.
D. Wherever is convenient for you.

27. You should learn to stop your vehicle ...
A. Before stop lines or crosswalks.
B. At the stop lines or crosswalks.
C. After the stop lines or crosswalks.
D. At a convenient place.

28. Good driving habits include ...
A. Identifying obstructions and avoiding them.
B. Stopping your vehicle at a convenient location.
C. Driving behind obstructions and hoping for the best.
D. Driving as you feel comfortable.

29. Another good driving habit you should cultivate is ...
A. Looking sideways only when driving.
B. Observing your environment thoroughly.
C. Depending exclusively on your mirrors for guidance.
D. All of the above.

30. The benefits of paying attention to your environment include all of the following except:
A. It makes it easier for you to spot potential problems.
B. You can avoid obstacles that may impede your driving.
C. It prevents you from colliding with other vehicles.
D. None of the above.

31. When changing lanes, which of the following habits is a part of safe driving?
A. Looking at your mirrors only.
B. Looking over your shoulders only.
C. Looking over your shoulders and your mirrors.
D. Changing lanes whenever you need to.

32. Speed limits are usually posted beside roads and elsewhere for what purpose?
A. To enable drivers to drive at a safe speed.
B. To enable the authorities to show off their powers.
C. They are not usually posted.
D. None of the above.

33. Abiding by the speed limit ...
A. Is proof of your safety consciousness.
B. Shows you are incompetent at driving at faster speeds.
C. Hampers your driving skills on highways.
D. Slows your driving.

34. When driving in bad weather, what takes precedence?
A. Speed limits take precedence over the bad weather.
B. The bad weather takes precedence over speed limits.
C. They take precedence over each other.
D. They are both immaterial.

35. Besides bad weather, another reason for driving within the speed limit is ...
A. The nature of the road.
B. The type of vehicle you drive.
C. The visibility.
D. The condition of your tires.

36. Anticipating other drivers' actions ...
A. Is only important for professional drivers.
B. Is a test of your ability to deal with a wide range of drivers.
C. Allows you to test how powerful your vehicle's horn is.
D. Doesn't impact your driving.

37. What is the purpose of pavement markings?
A. They convey important pieces of information to road users.
B. They are designed to beautify roads.
C. They indicate the beginning and end of a road.
D. They have different meanings according to the prevailing circumstances.

38. Yellow lines on pavement serve what purpose?
A. They serve as traffic separators.
B. They serve as traffic warnings.
C. They serve as security indicators.
D. They serve as warnings about potential obstacles.

39. A solid yellow line means what?
A. It indicates the part of the road designed for only cyclists.
B. It indicates the part of the road closed to the public for an event.
C. It indicates a part of the road undergoing renovation.
D. It indicates parts of the road only meant for use in winter.

40. Yellow dashes mean what?
A. The road is closed temporarily.
B. The road is open for use.
C. The road is for emergency use only.
D. The road is under construction.

41. A minimum of ... points deducted is acceptable during a road skills test.
A. 30.
B. 20.
C. 10.
D. 50.

42. Getting involved in an accident during a road skills test will result in ...
A. A $500 fine.
B. Automatic disqualification.
C. Automatic disqualification with a $200 fine.
D. Automatic disqualification and a learner's permit revocation.

43. A B Class restriction on a driver's license indicates that ...
A. The driver can only operate a vehicle with the assistance of corrective lenses.
B. The driver is eligible to operate tankers.
C. The driver can transport hazardous substances.
D. The driver is limited to 18 passengers on their passenger bus.

44. A C Restriction on your license indicates that...
A. You can only drive a vehicle with a mechanical aid.
B. You can only drive a vehicle with a hearing aid.
C. You can drive a wide range of commercial motor vehicles.
D. None of the above.

45. An effective communication channel between drivers is the appropriate use of ...
A. Indicators.
B. Blinkers.
C. Hand signals.
D. Brake lights.

46. An E Restriction limits your driving to ...
A. Commercial vehicles with an automatic transmission only.
B. Commercial vehicles with a manual transmission only.
C. Hybrid vehicles only.
D. None of the above.

47. If you are restricted from night driving, you will be issued a license with a ...
A. G Restriction.
B. B Restriction.
C. F Restriction.
D. H Restriction.

48. The M Restriction restricts you from operating
A. A Class A passenger vehicle or school bus under certain conditions.
B. A Class B vehicle under certain conditions.
C. A Class C vehicle under certain conditions.
D. All classes of vehicle under certain conditions.

49. A general condition for overriding any of these restrictions is ...
A. Holding a restriction-free Commercial Learner's Permit for 14 days.
B. Holding a restriction-free Commercial Learner's Permit for 24 days.
C. Holding a restriction-free Commercial Learner's Permit for 20 days.
D. Holding a restriction-free Commercial Learner's Permit for 10 days.

50. If your driver's license carries an O Restriction, you can't operate ...

A. Class A vehicles.

B. Class B vehicles.

C. Class C vehicles.

D. Tractor trailer with a fifth-wheel connection.

34 Air Brakes Test

1. What happens when a vehicle equipped with air brakes is idle?
A. The vehicle will lose its air pressure.
B. The contaminants can settle easily.
C. The vehicle's air system will be charged.
D. Nothing will happen to the vehicle.

2. What effect does depressing the brake pedal have on the pressure?
A. The pressure remains stationary.
B. The pressure increases gradually.
C. The pressure decreases gradually.
D. The pressure evaporates gradually.

3. Which parts of the air brakes support the compressor?
A. The emergency air brakes.
B. The compressor governor.
C. The air tank drains.
D. None of the above.

4. What must you do before driving a vehicle equipped with the air brake system?
A. Empty the air drain tank.
B. Push the emergency brake.
C. Depress the brakes.
D. Test the horns and the blinkers.

5. What produces the force applied to a vehicle's wheels for effective braking?
A. Compressed air.
B. Mechanical force produced by the brake system.
C. The brake pedal.
D. The air tank governor.

6. What occurs when the brake pedal is pressed?
A. The compressed air is released.
B. The compressed air is compressed further.
C. The compressed air is circulated through the braking system.
D. The compressed air becomes hot.

7. The air brake system is a combination of ...
A. Four braking systems.
B. Five braking systems.
C. Two braking systems.
D. Three braking systems.

8. This braking system applies and releases the brakes when you apply the brakes while driving normally.
A. The Parking Brake System.
B. The Service Brake System.
C. The Control Brake System.
D. The Automatic Brake System.

9. To use the ..., its air must be routed under pressure at approximately 100 to 120 psi.
A. Parking Brake System.
B. Service Brake System.
C. Control Brake System.
D. Automatic Brake System.

10. This brake is otherwise known as a hand brake or e-brake ...
A. The Service Brake System.
B. The Parking Brake System.
C. The Emergency Brake System.
D. The Automatic Brake System.

11. How do heavy vehicle drivers know whether there is enough pressure in the brake system or not?
A. Through the air pressure meter.
B. Through a built-in gauge.
C. Through an air pressure compressor governor.
D. None of the above.

12. What is a wig wag?
A. A device that indicates when the brake is faulty.
B. A device that indicates that the air pressure is below the recommended level.
C. A device for measuring brake efficiency.
D. A device for fixing faulty brakes.

13. The air pressure warning signal goes off when ...
A. The pressure in the air pressure tank is below 60 psi.
B. When the pressure in the tank is below 50 psi.
C. When the pressure in the tank is below 100 psi.
D. When the pressure in the tank is below 45 psi.

14. Under what conditions can the warning signal go off when the pressure is between 80 and 85 psi?
A. When driving a large bus.
B. When driving a trailer tractor.
C. When driving a passenger bus.
D. When driving a tanker.

15. The compressed air brake system is classified into ...
A. Three classes.
B. Four classes.
C. Two classes.
D. Five classes.

16. Two service brake circuits, a parking brake circuit and a trailer brake circuit are some components of the ...
A. Circuit system.
B. Brake system.
C. Supply system.
D. Control system.

17. A brake pedal air valve regulates the ... and ...
A. Front and rear wheel circuits.
B. Parking brake and trailer brake circuits.
C. Parking and wheel circuits.
D. Front and trailer brake circuits.

18. Which of these components does the engine use to drive the air compressor?
A. A crankshaft pulley or a belt.
B. A crankshaft pulley or an iron chain.
C. An iron chain or a crankshaft pulley.
D. None of the above.

19. What must you do first to test your low pressure warning signal?
A. Apply the brake first.
B. Turn off the engine.
C. Check the air compressor.
D. Reduce the pressure in the air tank.

20. Under what conditions should you test your vehicle's air leakage rate?
A. When the system is half charged.
B. When the air pressure is between 200 and 250 psi.
C. When the system is fully charged.
D. When the vehicle's engine has been run continuously for five hours.

21. The recommended operating pressure per square inch for a bus' brake system is ...
A. 60 pounds per square inch.
B. 80 pounds per square inch.
C. 85 pounds per square inch.
D. 50 pounds per square inch.

22. A reasonable air leakage rate for combination vehicles is ...
A. No more than 8 psi.
B. More than 5 psi.
C. No more than 4 psi.
D. 10 psi.

23. Under what conditions should you be concerned about the cracks in the brake drums or discs?
A. If the cracks are less than one inch long.
B. If the cracks are shorter than half the width of the areas allowed for friction.
C. If the cracks are longer than half the width of the areas allowed for friction.
D. It depends on the type of vehicle.

24. When driving a vehicle equipped with a dual air brake system, you should always ...
A. Allow sufficient time for the air pressure in the air compressor to build up.
B. Allow sufficient time for the air pressure in the air compressor to be reduced drastically.
C. Allow sufficient time for the air pressure in the air compressor to rise somewhat.
D. Do nothing.

25. The dual air brake is divided into the ... and ...
A. Front and rear system.
B. Primary and secondary systems.
C. Manual and automatic system.
D. Stop and start system.

35 Doubles and Triples Test

1. How do you prevent the rig from moving?
A. By using the emergency brakes.
B. By using the parking brakes.
C. By using both the emergency and parking brakes.
D. All of the above.

2. When uncoupling the converter dolly, what comes first?
A. Lowering the landing gear.
B. Connecting safety chains.
C. Releasing the pintle hook.
D. Moving away from the dolly slowly.

3. When is it unwise to detach or unlock the pintle hook?
A. After lowering the landing gear.
B. Before chocking the wheels.
C. After releasing the pintle hook.
D. When the dolly is under the rear trailer.

4. When coupling a triple trailer, which of these parts must be coupled first?
A. The first and second trailers.
B. The third and fourth trailers.
C. The second and third trailers.
D. It depends on the type of vehicle.

5. The first thing you must do when uncoupling a triple trailer is...
A. Uncouple the second and third trailer first.
B. Uncouple the first and second trailer.
C. Pull the dolly out and unhitch it.
D. Unhitch the third trailer first.

6. During an inspection, which part of the double or triple trailer must you check first?
A. The lubrication.
B. The locking jaws.
C. The lower section of the wheel.
D. The upper fifth wheel.

7. Spaces between the lower and upper fifth wheel are a sign of what problem?
A. The fifth wheel is not properly connected.
B. A bolt has worn out.
C. It is not properly lubricated.
D. A part is missing.

8. What should you do when in doubt about the best way to couple or uncouple a trailer?
A. Use your discretion.
B. Check the manufacturer's handbook.
C. All of the above.
D. None of the above.

9. When you are checking the lower part of the fifth wheel, you should focus on all of the following except:
A. The lubrication.
B. The locking jaws.
C. The release arm.
D. None of the above.

10. The glide plate should be mounted...
A. On the trailer rear.
B. On the trailer frame.
C. On the fifth wheel.
D. Beside the kingpin.

11. Why is it important to check the distance between the fifth wheel and the tractor frame?
A. To ensure they avoid hitting each other during turns.
B. To ensure the connection is solid.
C. To ensure that nothing can get between them and interfere with their operations.
D. It's a preventive measure against disconnecting when the vehicle is in motion.

12. For the best performance, the landing gear must be...
A. Reasonably raised.
B. Partially raised.
C. Fully raised.
D. Not raised.

13. The tractor air supply control is otherwise known as a...
A. Trailer emergency valve.
B. Tractor protection valve control.
C. Tractor emergency valve control.
D. Both A and B.

14. Check all of the following in the trailer emergency brakes except the:
A. Emergency position.
B. Air pressure.
C. Brake parts.
D. None of the above.

15. When testing airflow to the trailer, it is important that you first...
A. Chock the wheels or apply the brakes.
B. Wait until the air pressure is normal.
C. Push the trailer air supply knob.
D. Use an air pressure meter to read the current air pressure.

16. The last step when testing airflow to the trailer is...
A. Supplying air to the service line.
B. Pushing the trailer air supply knob.
C. Waiting until the air pressure normalizes.
D. Reading the air pressure with a meter.

17. This type of brake is needed for intermittent testing:
A. An emergency brake system.
B. A trailer service brake.
C. A hand brake.
D. An e-brake.

18. All of the following parts of a trailer service brake need to be thoroughly checked except:
A. The air pressure.
B. The air pressure meter.
C. The brake's reliability.
D. Both A and C.

19. You can protect the tractor air brake system with the...
A. Air brake protector.
B. Trailer brake guard.
C. Tractor protection valve.
D. Automated parking brake protector.

20. At what air pressure range should you stop airflow to the trailer?
A. 40 to 50 psi.
B. 50 to 60 psi.
C. 20 to 45 psi.
D. 33 to 40 psi.

21. When checking the power-operated landing gear, you should always make sure to check...
A. The power button.
B. The landing gear and power connector.
C. Air leaks.
D. Landing gear protector.

22. How do you remove some of the weight on a dolly?
A. By offloading the vehicle.
B. By removing the landing gear.
C. By lowering the second trailer's landing gear.
D. By detaching the second trailer.

23. How do you check the second trailer for air?
A. Through the emergency line shutoff.
B. Through the air meter.
C. By lifting the valve over it.
D. By increasing the air pressure level.

24. The crack-the-whip effect occurs mostly in the... of a combination vehicle.
A. The tractor.
B. The connecting section.
C. The trailer.
D. A and C.

25. As a double or triple trailer driver, are you allowed to operate in all US states?
A. Yes.
B. The DMV must decide.
C. It depends on the years of experience and the driver's age.
D. No.

36 Hazmat Test

1. All of the following are endorsement requirements for transporting hazardous materials except:
A. A valid CDL permit.
B. Proof of identity.
C. A valid certificate from a medical examiner.
D. Proof of residency.

2. Which of these is acceptable proof of identity for a US citizen?
A. Certificate of US Citizenship.
B. A US passport.
C. Certificate of naturalization.
D. All of the above.

3. There are ... classes of hazmat endorsement.
A. Two.
B. Five.
C. Four.
D. Three.

4. This hazmat endorsement type is required for driving tankers...
A. P Endorsement.
B. K Endorsement.
C. N Endorsement.
D. X Endorsement.

5. This is considered the ultimate endorsement class...
A. X Endorsement.
B. H Endorsement.
C. V Endorsement.
D. Y Endorsement.

6. A hazmat endorsement on your Commercial Driver's License will cost you...
A. $100.
B. $150.
C. $50.
D. $200.

7. You can get a discount if you are...
A. A new applicant with a Commercial Learner's Permit.
B. A new applicant with a valid TWIC card.
C. A new applicant with five years' driving experience.
D. A new applicant without a criminal record.

8. What is a TWIC?
A. Transport Worker Identification Card.
B. Transport Worker Identification Column Card.
C. Transportation Worker Identification Credential.
D. None of the above.

9. A hazmat endorsement is valid for...
A. Ten years.
B. Fifteen years.
C. Five years.
D. Four years.

10. Which of the following information is not required on a TSA application?
A. Middle name.
B. Date of birth.
C. Preferred method of contact.
D. None of the above.

11. The most important pieces of information on the TSA application are:
A. Phone number and email address.
B. Name and contact method.
C. Date of birth and zip code.
D. Gender and middle name.

12. Which of the following is/are not allowed in the name fields on the application form?
A. Apostrophe.
B. Hyphens.
C. Letters.
D. Special characters.

13. Do you have to take the assessment in a certain place?
A. Yes.
B. It depends on the examiner.
C. No.
D. It depends on you.

14. You need all of the following information to choose a location except:
A. The test you want to take.
B. Your zip code.
C. Your city.
D. Your credit history.

15. The passing grade for the hazmat endorsement test is...
A. 74%.
B. 85%.
C. 90%.
D. 80%.

16. You are allowed to retake the exam...
A. As many times as possible.
B. 8 times.
C. Less than 5 times.
D. Up to 10 times.

17. Is retaking the test free?
A. Yes.
B. No.
C. It is free for the first two times but you must pay for subsequent retakes.
D. You must pay for the first two retakes while subsequent ones are free.

18. The assessment test will assess your knowledge of all of the following except:
A. Your ability to identify hazardous materials.
B. Your ability to placard your vehicle according to established rules.
C. Your ability to load shipments safely.
D. None of the above.

19. The vessels that an applicant with TWIC card has access to are regulated by the...
A. Maritime Transportation Security Act of 2010.
B. Maritime Transportation Security Act of 2000.
C. Maritime Transportation Security Act of 2015.
D. Maritime Transportation Security Act of 2002.

20. What makes up the cost of a hazmat endorsement?
A. TSA charges plus state charges.
B. State charges plus federal charges.
C. TSA charges plus federal charges.
D. TSA, state, and federal charges.

21. Drivers who transport placarded hazardous materials need a...
A. C Endorsement.
B. P Endorsement.
C. M Endorsement.
D. H Endorsement.

22. The DOT Card is otherwise known as a...
A. Medical History Certificate.
B. Medical Health Record.
C. Medical Examiner's Credentials.
D. Medical Examiner's Certificate.

23. Hazmat endorsements are classified into...
A. H Endorsement, P Endorsement, and K Endorsement.
B. P Endorsement, Y Endorsement, and V Endorsement.
C. H Endorsement, X Endorsement, and N Endorsement.
D. V Endorsement, F Endorsement, and E Endorsement.

24. If your waiver or appeal is granted, you and the...
A. State government.
B. Federal government.
C. Local Department of Motor Vehicles.
D. All of the above.

25. All of the following are ways to identify hazardous materials except...
A. A highlighted entry.
B. An X or H endorsement.
C. Placarding.
D. None of the above.

37 Tanker Test

1. A space left on top of a tank during loading to accommodate expansion is called...
A. An outage.
B. A surge allowance.
C. A bulkhead corrector.
D. Expansion room.

2. A very important attribute of a liquid each driver should know prior to loading is...
A. Its viscosity.
B. Its capacity volume.
C. Its expansion rate.
D. Its boiling and freezing points.

3. What are baffles?
A. Internal structures in a tank for storing extra liquids.
B. Internal structures in a tank with some holes.
C. Equipment that is specially designed for surge prevention.
D. None of the above.

4. Baffles are necessary in tanks for...
A. Bulkhead prevention.
B. Surge prevention.
C. Rollover prevention.
D. Accident prevention.

5. What is the difference between baffles and smooth bore tanks?
A. Baffles have internal partitions and smooth bore tanks don't.
B. Smooth bore tanks have internal partitions but baffles don't.
C. Smooth bore tanks are larger than baffles.
D. Baffles are larger than smooth bore tanks.

6. Which of these tanks is used for transporting foods and food products?
A. Smooth bore tanks.
B. Baffles.
C. Tanks with outage.
D. Tanks with internal partitions.

7. When driving on wet roads, it is advisable to...
A. Double the stopping distance.
B. Stop at your discretion.
C. Reduce your speed considerably.
D. Both A and C.

8. A safety precaution when entering a tunnel is to...
A. Blast your horn repeatedly.
B. Accelerate a bit.
C. Remove your sunglasses.
D. All of the above.

9. Are tanker drivers legally obligated to obey speed limits when driving around corners?
A. No.
B. Yes.
C. It's at their discretion.
D. The law varies from state to state.

10. Who is responsible for cleaning up tanker spills?
A. The Environmental Protection Agency.
B. The Department of Transportation.
C. The tanker driver.
D. The Department of Road Safety.

11. One of the effects of quick steering is...
A. Rollover.
B. Surging.
C. Bulkheads.
D. Brake failure.

12. When a tire slips on the road, this is known as...
A. Skidding.
B. Overaccelerating.
C. Sliding.
D. Relative jackknifing.

13. The first corrective measure for skidding is to...
A. Release the brake after stopping.
B. Release the brake midway through the skidding.
C. Release the brake immediately.
D. Release the brake at your convenience.

14. Which of the following areas should you check for leaks?
A. Under the vehicle.
B. The tires.
C. The manhole covers.
D. All of the above.

15. Covers and ports must have...
A. Gaskets.
B. Manholes.
C. Discharge ports.
D. Hole sprays.

16. Some equipment all drivers must have includes...
A. Built-in fire extinguishers.
B. Grounding or bonding cables.
C. Vapor recovery kits.
D. All of the above.

17. Electrical accessories are best connected with...
A. Electrical cables.
B. Bonding or grounding cables.
C. Electrical connectors.
D. Electrical connecting devices.

18. Other parts of the vehicle you should inspect include all of the following except...
A. Hoses.
B. Cut-off valves.
C. Vents.
D. None of the above.

19. The Tanker Endorsement Knowledge Test costs...
A. Between $10 and $50.
B. $50.
C. $100.
D. $80.

20. When applying brakes, the pressure should be...
A. Increasing.
B. Decreasing.
C. Steady.
D. Fluctuating.

21. It is advisable to... when driving through a curve.
A. Slow down.
B. Accelerate significantly.
C. Accelerate slightly.
D. Slow down a little and increase speed.

22. All of the following can cause skidding in tankers except:
A. Quick steering.
B. Overaccelerating.
C. Excessive braking.
D. Checking the mirror for incoming vehicles.

23. A complete guide for tanker inspection can be found in/on...
A. YouTube.
B. Google.
C. Bing.
D. The manufacturer's manual.

24. A vapor recovery kit is important...
A. To keep the vehicle clean.
B. To reduce pollution.
C. To ensure the vapor is free of contaminants.
D. All of the above.

25. All drivers must always carry...
A. Personal Protection Equipment.
B. Personal Protective Equipment.
C. Personal Vehicle Protective Equipment.
D. Personal Vehicle Protection Equipment.

38 Passenger Transport Test

1. An inspection under the vehicle should include:
A. The driveshaft.
B. The frame.
C. The exhaust system.
D. All of the above.

2. What is the appropriate speed for testing the service brake?
A. 5 miles/hour.
B. 4 miles/hour.
C. 10 miles/hour.
D. 20 miles/hour.

3. Are stickers appropriate on the windshield?
A. Yes.
B. No.
C. It depends on where a sticker is located.
D. It's up to a driver's discretion.

4. What are spare breakers or fuses used for?
A. They are used for preventing damage to the fuses or breakers.
B. They are used to replace blown fuses or breakers.
C. They are used to prevent leaks.
D. None of the above.

5. A good fire extinguisher must be... and...
A. Partially charged and mounted.
B. Mounted and fully charged.
C. Red and fully charged.
D. Partially charged and lightweight.

6. Triangles must be...
A. Blue and reflective.
B. Red and opaque.
C. Red and reflective.
D. Green and reflective.

7. The aisle of a passenger vehicle must always be...
A. Clear and encourage free traffic flow.
B. Allow extra space for physically challenged passengers.
C. Be used at the driver's discretion.
D. All of the above.

8. What are No-Zone areas?
A. Parts of the road closed to heavy vehicles.
B. Parts of the road closed to passenger buses.
C. Blind spots with the potential for triggering crashes.
D. Parts of the road where you can't pick up or drop off passengers.

9. Driving slowly is required when driving near...
A. Work areas and gas stations.
B. No-Zone areas and gas stations.
C. Work areas and No-Zone areas.
D. Closed buildings and gas stations.

10. Crashes are more frequent... on construction sites.
A. Later in the day.
B. At night.
C. During the day.
D. In the afternoon.

11. Two-thirds of traffic fatalities across the United States are caused by...
A. Driving under the influence of drugs or alcohol.
B. Defensive driving.
C. Aggressive driving.
D. A and C.

12. ... and ... are two major characteristics of defensive driving.
A. Keeping a safe distance between vehicles and maintaining a safe speed.
B. Preventing other motors from overtaking you and blasting the horn.
C. Driving on the right side of the road and playing loud music.
D. Driving with the right license and endorsement.

13. ...is/are arguably one of the most important safety precautions for drivers.
A. Regular and thorough inspections.
B. Knowing how to drive on slippery roads.
C. Fastening seat belts while driving.
D. Applying the brakes at regular intervals.

14. Passenger buses are allowed to carry what class of ammunition?
A. Special military ammunition.
B. Ordinary ammunition.
C. Made in Korea ammunition.
D. Small-arms ammunition.

15. Leaving baggage in the doorway can lead to... and...
A. Obstruction and theft of the baggage.
B. Obstruction and damage to the baggage.
C. Damage to the doorway and the baggage.
D. A and C.

16. Passenger vehicles cannot carry any of the following except:
A. Class 6 poison.
B. Division 2.3 poison gas.
C. Radioactive materials in spaces meant for passengers.
D. None of the above.

17. A passenger vehicle can transport all of the following except:
A. Emergency hospital supplies.
B. ORM-D.
C. Small arms ammunition.
D. Class 6 poison.

18. Under what conditions can a passenger vehicle transport animals?
A. If they are owned by the driver and are being taken to a vet.
B. If they are homeless and need serious medical attention.
C. If they are for certified service or serve as guides.
D. If they are hit by a hit-and-run driver and need urgent medical assistance.

19. A charter bus driver should only allow passengers in the vehicle...
A. When embarking on a trip.
B. As soon as they arrive.
C. 30 minutes before the trip.
D. As determined by the driver.

20. What are the three triangles?

A. A set of triangles with three different colors.

B. A set of triangles you can put beside the road as a warning signal to other drivers of a crash ahead.

C. Equal-sized triangles used in construction sites and passenger buses.

D. A special type of triangle that serves different purposes in a vehicle.

21. What is a driveshaft?

A. Another name for the gear.

B. A device used for connecting the components of a drive train in a large vehicle.

C. A special component specifically designed for passenger vehicles.

D. A special driver for some components of the passenger bus.

22. The following are signs of a faulty driveshaft:

A. Intense vibration underneath the vehicle.

B. Shuddering while accelerating.

C. Turning problems.

D. All of the above.

23. A common problem you should look out for in a vehicle's frame is...

A. Cracks and loose bolts.

B. Loose bolts and improper lubrication.

C. Friction and cracks.

D. Friction and loose bolts.

24. Abnormal noise in a driveshaft is caused by...

A. Improper lubrication.

B. Failed or worn-out supporting bushings or bearings.

C. Cracks or holes in the driveshaft.

D. A deteriorating shaft handler.

25. These are all common problems in driveshafts except...

A. Shuddering during acceleration.

B. Turning problems.

C. Too much noise in the driveshaft.

D. None of the above.

39 School Bus Test

1. The first step towards handling an emergency is...
A. Recognizing the problem.
B. Seeking help.
C. Fixing the problem.
D. Shifting the blame on other road users.

2. What factor must you first consider when determining whether or not to evacuate students?
A. Are they in greater danger if they remain on the school bus?
B. How will their parents respond to the issue?
C. What is the most appropriate time to notify the relevant authorities?
D. What is the best place to move the students?

3. Under what conditions is evacuation mandatory?
A. Higher possibility of accident and a threat of fire.
B. If the bus changes position or hazardous materials spills around or on the bus.
C. None of the above.
D. All of the above.

4. When planning ahead for emergencies that may require student evacuation, you need the assistance of... and...
A. Tall and agile students.
B. Responsible and older students.
C. Tall and older students.
D. Responsible and final year students.

5. It's important to... prior to an evacuation.
A. Explain the evacuation procedures to the students.
B. Take a roll call of the students.
C. Organize students according to their needs.
D. Call the relevant authorities.

6. When students are at risk from spilled hazardous materials...
A. Lead them downwind of the bus.
B. Keep them in the vehicle with the windows and doors closed and locked.
C. Lead them upwind of the bus.
D. Reassure them of their safety.

7. When an emergency happens close to railroad tracks, you must...
A. Lead the students towards the railroad tracks for assistance.
B. Lead the students away from the tracks.
C. Lead the students away from the direction of oncoming trains.
D. Call officials for assistance.

8. How do you secure a vehicle during emergencies?
A. Shut the engine off and place the vehicle in neutral.
B. Leave the ignition key in and activate the hazard lights.
C. Keep the engine running and place the three triangles beside the vehicle.
D. All of the above.

9. Which of these classes of students shouldn't be evacuated immediately?
A. Older students.
B. Students with head or neck injuries.
C. Ill students.
D. Foreign students.

10. You can ask all of the following people for assistance except:
A. Passing motorists.
B. Passing pedestrians.
C. Responsible students.
D. None of the above.

11. As a school bus driver, your knowledge should include:
A. Accident prevention.
B. Accident management.
C. Accident prevention and management.
D. Accident analysis, prevention and management.

12. You shouldn't release the students to any of the following people except:
A. Parents.
B. Neighbors.
C. Other family members.
D. None of the above.

13. The student evacuation process should begin with...
A. A call to the bus dispatch.
B. A call to the emergency team.
C. A visit to the nearby hospital.
D. A call to your supervisor or the school authorities.

14. Some evacuation options you can choose from include:
A. Roof hatch.
B. Side doors.
C. Rear and front door.
D. All of the above.

15. Steer clear of the bus until it is declared fit for use by...
A. The police and accident management team.
B. Bus dispatch and the relevant authorities.
C. Bus dispatch and road safety management team.
D. School authorities and the police.

16. When emergencies are caused by inclement weather, it is best to...
A. Keep the students in the vehicle until the weather improves.
B. Keep the students in the vehicle until a rescue team arrives.
C. Evacuate the students to a safe assembly area.
D. Evacuate the student to a larger vehicle for immediate transportation to their destination.

17. What is a school lockdown?
A. This refers to locking up a school during the holidays.
B. This refers to a school closure triggered by power failure.
C. This refers to the act of barring people from exiting or entering a school until further notice.
D. None of the above.

18. Some common factors that can trigger a lockdown include all of the following except:
A. Natural disasters.
B. Police activity in the neighborhood.
C. Medical emergencies.
D. All of the above.

19. There are... types of lockdown.
A. Four.
B. Three.
C. Ten.
D. Two.

20. Medical emergencies, weather-related emergencies, and traffic jams are typical examples of a...
A. Code Black Lockdown.
B. Code Yellow Lockdown.
C. Code Red Lockdown.
D. Code Green Lockdown.

21. Gunshots, a hostage situation, and the presence of a dangerous person in or around the school are some typical examples of a...
A. Code Black Lockdown.
B. Code Yellow Lockdown.
C. Code Red Lockdown.
D. Code Green Lockdown.

22. What is a shelter-in-place?
A. This occurs when the school bus is contaminated with hazardous materials.
B. It is a situation that may arise when the school environment is contaminated with hazardous materials.
C. It refers to a location where students can be kept safe during a lockdown.
D. It is a place for cleaning the school bus when contaminated with hazardous materials.

23. What should you do when your vehicle develops mechanical trouble?
A. Drive it to a safe place before stopping.
B. Secure it and activate the four-way flashers.
C. All of the above.
D. None of the above.

24. The Antilock Brake System that is designed to minimize skidding in school buses is...
A. Electronic controller.
B. Wheel speed sensors.
C. Modulators.
D. A and B.

25. How many times are you allowed to retake the test per permit?
A. Three times.
B. Five times.
C. Ten times.
D. Two times.

33 Answers for General Knowledge Test

(1) (A) Department of Motor Vehicles.

You can schedule a time to take the test by visiting the official website of the Department of Motor Vehicles. On the website, you will find a practical guide on how to proceed with the test, the requirements and other pieces of information you will find invaluable to performing well on the test.

(2) (B) Saturday.

When planning your road test, you can only schedule an appointment on business days—Monday through Friday.

(3) (A) 30 minutes.

It is advisable that you arrive at the center at least 30 minutes in advance in order to account for traffic, weather, etc.

(4) (C) Rhode Island only

Applicants can only schedule an appointment for the driver's license test in the states of their residence. Hence, a Rhode Island resident can register for the exam and take it in that state only, not elsewhere. The same applies to other drivers across the country.

(5) (B) Official website of the local Department of Motor Vehicles.

If you need a comprehensive list of accredited testing centers, visit the official website of the local Department of Motor Vehicles.

(6) (C) Two times.

If you fail the test on the first attempt, you can still take the exam two times on your permit. The rule allows each permit holder to take the exam a maximum of three times. This is to enable drivers to prepare in advance for the test.

(7) (C) If you are physically unfit to operate a vehicle.

If you exhaust the three chances allocated to a driver on a learner's permit, you can apply for another permit and have three more chances to attempt the test. However, your permission to apply for the license becomes null and void if you are physically unfit to operate a vehicle.

(8) (A) Seven days.

If you are under 28 and fail the test, you must allow a seven-day interval before reapplying to take the exam.

(9) (B). Go through your study materials again and improve your driving skills.

While waiting to submit your retake application, keep studying and practicing your driving. Make good use of your time in between tests.

(10) (A) Cancel your first appointment.
Before you are allowed to reschedule your driver's license test, it is mandatory that you cancel your first appointment.

(11) (A) No.
As a rule, no driver is allowed to hold two testing appointments simultaneously. Hence, if you are not satisfied with your first appointment for whatever reason, cancel it first and schedule a second appointment.

(12) (A) Change of job and inclement weather.
There are several reasons why you may consider rescheduling your test appointment. Some of these reasons included a change of job and inclement weather. If you are late to the examination center, that's your fault. It is not a genuine reason for changing your test date.

(13) (D) None of the above.
Running a red light, driving on curbs and cutting corners when driving on winding roads are some of the bad driving habits that may cost you your license. If you engage in any of these habits while taking the test, your examiner will score you low.

(14) (A) Maintain your cool no matter what.
During the test, you will have to contend with different challenges such as reckless drivers who don't care about other road users. When another driver nearly gets you involved in an accident, don't lose your cool. Your reaction will tell an examiner a lot about you.

(15) (C) All of the above.
You have a better chance of passing the examination if you keep calm before the test. Also avoid drugs and alcohol which will impair your judgment behind the wheel.

(16) (B) 30 days.
To qualify for the road skills test, you must have a valid CDL for a minimum of 30 days before scheduling the test.

(17) (B) The Commercial Driver Road Test Lot.
As a prerequisite for obtaining the license, visit the Commercial Driver Road Test Lot.

(18) (B) 15 minutes.
When taking any test, including the road skills test, make sure you always arrive at least 15 minutes early to the testing center.

(19) (B) 40-hour Certification of Eligibility for Provisional License form.

If you are under 21 and taking the road test, you must provide proof of your driving ability and skills. This includes a 40-hour Certification of Eligibility for Provisional License form.

(20) (B) A licensed driver.

Each license applicant taking the test is expected to be accompanied by a licensed driver, especially if you drive yourself to the center. This is to ensure that you have someone to guide you as you drive to the test center and back home.

(21) (B). Banned from taking the test for six months.

If you drive yourself to the test center without a licensed driver to guide you, you stand the risk of being banned from taking the test for six months.

(22) (B) 15 hours.

While honing your driving skills, you are expected to learn how to drive a commercial vehicle of your choice under different conditions. This includes learning how to drive at night.

(23) (C) 10 hours.

Your driving practice should cover areas of moderate to heavy traffic as well. You are required to dedicate at least 10 hours to driving under such conditions. The more you practice, the better you become.

(24) (B) Shortly before reaching the stopping position.

You should avoid jerky braking. To do this, stop the vehicle and bring it to a rest shortly before reaching the stopping position.

(25) (B) Grinding.

Grinding is a common error among new and inexperienced drivers. To pass the test, you must avoid committing this error at all costs.

(26) (A) Before the intersection.

When you are approaching an intersection and your examiner asks you to stop, you must stop the vehicle before the intersection. This enables you to have a clear view of the intersection whenever you pull out.

(27) (A) Before stop lines or crosswalks.

When approaching crosswalks or stop lines, stopping before them is a good driving habit for safety's sake.

(28) (A) Identifying obstructions and avoiding them.

Good driving habits ensure your safety on the road. They guarantee the safety of other road users as well. Hence, endeavor to identify obstructions when driving. As much as possible, avoid them at all costs. Don't deliberately put yourself in situations that make driving more difficult for you.

(29) (B) Observing your environment thoroughly.
The road ahead shouldn't be your only concern when you are behind the wheel. Take your environment into consideration too. Look sideways when driving to spot potential problems and avoid them. Check your mirrors regularly as well. You can't ever be too careful. Do whatever it takes to be safe.

(30) (D) None of the above.
You benefit a lot if you pay attention to your environment when you are behind the wheel. Aside from making it easier for you to identify potential problems and avoid them, it also enables you to drive with ease. Your chances of running into other vehicles are reduced too.

(31) (C) Looking over your shoulders and your mirrors.
When you wish to change lanes, don't assume that other road users are aware of your intentions. That may land you in trouble. Rather, look over your shoulders to check the vehicles behind you. Check your mirrors for oncoming vehicles from other sides too.

(32) (A) To enable drivers to drive at a safe speed.
When driving, you may see speed limits posted. They serve as reminders to drive at a safe speed for your sake and that of others.

(33) (A) It's proof of your safety consciousness.
Abiding by the speed limits is proof of your safety consciousness. If you violate them, you are invariably showing your lack of respect for your safety and the safety of others.

(34) (B) The bad weather takes precedence over speed limits.
When driving in bad weather, you are not obligated to keep to the speed limit. The bad weather takes precedence. The rule of thumb is to drive at a speed that ensures everyone's safety.

(35) (C) The visibility.
Besides bad weather, another factor you should consider when considering whether to drive at the exact posted speed limit is visibility. Hitting 60 miles/hour if you can barely see is a bad idea. It may be safer to keep to 20 miles/hour until the weather clears.

(36) (B) A test of your ability to deal with a wide range of drivers.
While driving, it is dangerous to focus only on your own driving. You must consider other road users as well. Their driving decisions may affect you. Hence, anticipate their actions and make adequate preparations for them.

(37) (A). Markings that convey important pieces of information to road users.

Pavement markings are used to convey important messages to road users. Abiding by that information helps keep everyone on the road safe.

(38) (A). They serve as traffic separators.

Yellow lines serve as traffic separators so that vehicles going in opposite directions won't run into each other.

(39) (B). It indicates the part of the road closed to the public.

A solid yellow pavement marking indicates the part of the road closed to the public for a particular reason such as repairs being made.

(40) (B). The road is open for use.

While yellow dashes indicate you should steer clear, yellow markings mean the opposite. They indicate that part of the road is safe to use.

(41) (A). 30.

For each driving rule you violate during the road skills test, your examiner will deduct points. The acceptable limit is 30. A driver who has more than 30 points deducted during the test has failed. This underscores the importance of obeying every driving rule while taking the test.

(42) (B). Automatic disqualification.

If you get involved in an accident while taking the road skills test, you have automatically failed the exam.

(43) (A). The driver can only operate a vehicle with the assistance of corrective lenses.

If you have a B restriction on your driver's license, this implies that you are only licensed to operate a vehicle with the assistance of corrective lenses. Hence, if you must use corrective lenses, you will not be allowed to operate the vehicle without wearing them.

(44) (A). You can only drive a vehicle with a mechanical aid.

A C Restriction on your license indicates that you are only permitted to drive a vehicle with a mechanical aid. Failure to do so will result in consequences.

(45) (B). Blinkers.

An effective communication channel between drivers is the effective use of blinkers. They signal to drivers where you are planning to turn so that they can then stay out of the way and remain safe.

(46) (A). Commercial vehicles with an automatic transmission only.

You can only drive commercial vehicles with an automatic transmission if you have an E Restriction. Thus, if you wish to operate commercial vehicles with manual transmission systems, this restriction does not allow for that.

(47) (A). G Restriction.

Not every driver can handle night driving. Individuals who cannot see well at night or who simply don't drive well at night, for whatever reason, are issued a license with a G Restriction.

(48) (A). A Class A passenger vehicle or school bus under certain conditions.

An M Restriction is another form of restriction for drivers who drive school bus or class A passenger vehicles. Such drivers are prevented from operating school buses or passenger vehicles if the skills test for this special endorsement was done with a class B passenger vehicle.

(49) (A). Holding a restriction-free Commercial Learner's Permit for 14 days.

A general condition for overriding any of these restrictions is to hold a restriction-free Commercial Learner's Permit for 14 days. At that point, the license will be replaced with a license without the restriction, provided that you satisfy other conditions.

(50) (D). Tractor trailer with a fifth-wheel connection

The O restriction forbids drivers from operating commercial motor vehicle tractors of any sort. Thus, if this appears on your commercial learner's permit or driver's license, a tractor with a fifth-wheel connection is off the list of vehicles you can operate.

34 Answers for Air Brake Test

(1) (C). The vehicle's air system will be charged.

When a vehicle equipped with air brakes is parked, the vehicle's air system will be charged as the brake is not depressed. Then, the air pressure in the brake system will overcome the S-Cam or the diaphragm in a closed position. This will result in the brake system being released.

(2) (C). The pressure decreases gradually.

If you depress the brake, the pressure will decrease gradually. The decrease will turn the S-Cam while the brake shoes are simultaneously spread against the drum. This will cause the reservoir tanks to be filled by the compressor so that the air pressure increases again as the pedal retracts.

(3) (A). The emergency air brakes.

The emergency air brakes support the compressor. They work to enhance the performance of the air brake system by enabling drivers to stop their vehicles during emergencies. You can activate the system by pulling the emergency air brake button on the dash.

(4) (B). Push the emergency brake.

Before you drive a vehicle equipped with the air brake system, push its emergency brake in. That will give the air brake system enough air before you drive the vehicle. The emergency brake will remain in a free state once the emergency system is pressurized.

(5) (B). Mechanical force produced by the brake system.

The mechanical force produced by the brake system produces the force applied to a vehicle's wheels for effective braking. The mechanical force prevents accidental leakage of the air pressure that the vehicle needs for braking. Thus, the vehicle's brakes have what they need to function properly.

(6) (C). The compressed air is circulated through the braking system.

When the brake pedal is pressed, the compressed air in the system is circulated from one part of the braking system to another part. This makes braking possible as the entire system has air to function.

(7) (D). Three braking systems.

The air brake system is made up of three braking systems. These are the Service Brake System, the Parking Brake System and the Emergency Brake System. The combination of these components enables the driver to use the appropriate system to stop the vehicle under different conditions.

(8) (B). The Service Brake System.

The Service Brake System applies and releases the brakes when you apply the brakes under normal driving conditions. It is used for slow driving or stopping rather than for emergencies when the brake doesn't function properly.

(9) (B). The Service Brake System.

You can only use the Service Brake System by routing its air under pressure at approximately 100 to 120 psi to the vehicle's brake chamber. This will engage the brake and make the vehicle stop or slow down, as needed.

(10) (C). The Emergency Brake System.

The Emergency Brake System is otherwise known as an e-brake or hand brake. It is a braking system that allows you to stop your vehicle whenever you want and prevent unfortunate accidents. Aside from its use during emergencies, it can also be used to keep vehicles motionless when parked.

(11) (B). Through a built-in gauge.

Modern heavy vehicles are designed with a built-in gauge that monitors the air pressure in the brake. You may get a warning light or hear a sound if the air pressure is way below the recommended level. This enables you to know whether the system is faulty or working properly.

(12) (B). A device that indicates the air pressure is below the recommended level.

A wig wag is a device that drops down automatically to indicate the air pressure is below the recommended level. This enables the driver to see the low level and act promptly.

(13) (A). The pressure in the air pressure tank is below 60 psi.

If the pressure in the air pressure tank goes below 60 psi, the warning signal will go off. The warning indicates that the vehicle is not safe to drive at the moment.

(14) (A). When driving a large bus.

The air pressure warning signal goes off when the pressure in the tank goes below 60 psi. However, there is an exception. When driving a large bus, the signal will only go off when the pressure is between 80 and 85 psi.

(15) (C). Two classes.

The compressed air brake system is classified into the Control System and the Supply System. The former controls all the activities in the brake system while the latter comprises different checks, safety valves and a drain valve. It supplies everything the brake needs.

(16) (D). Control System.

Two service brakes, a parking brake circuit and a trailer brake circuit are some components of the Control System. It uses these components to control all the activities that go on in the brake system.

(17) (A). Front and rear wheel circuits.

A brake pedal air valve regulates the front and rear wheel circuits. These circuits receive compressed air from the different reservoirs to serve as a source of added safety for the vehicle, especially if there is an air leak. It helps to minimize the impact of any such leak.

(18) (A). A crankshaft pulley or a belt.

The vehicle's engine uses a crankshaft pulley or a belt to drive the air compressor. Vehicles use fuel to generate power. The power is distributed to different components of the vehicle through the mechanical and electrical systems. A crankshaft pulley is a major component for power distribution.

(19) (B). Turn off the engine.

The first step for a low pressure warning signal test is to turn off the engine. Then, make sure that the air pressure is enough so that the low pressure warning signal isn't already activated before you start the test. Then, turn the electrical power on to proceed.

(20) (C). When the system is fully charged.

You should only test the air leakage rate when the system is fully charged. That is about 125 psi. Release the parking brake after turning off the engine. Time the rate at which the air pressure drops. If the loss rate is more than 2 psi/minute for single vehicles, that's a problem.

(21) (C). 85 pounds psi.

The recommended operating pressure per square inch for a bus' brake system is 85. At that pressure, the brake system will function at optimal capacity and reduce the risk of brake failure and other related issues.

(22) (C). No more than 4 psi.

For combination vehicles, the reasonable air leakage rate mustn't be more than 4 psi. Otherwise, the leakage rate is too much and can damage the vehicle.

(23) (C). If the cracks are longer than half the width of the area allowed for friction.

Brake drums or discs have cracks by default. However, you should be concerned about these cracks if they are longer than half the width of the areas allowed for friction. The longer cracks may interfere with the operations of the drums or discs and undermine the brake system's efficiency.

(24) (A). Allow sufficient time for the air pressure in the air compressor to build up significantly.

When driving a vehicle that is equipped with a dual air brake system, allow sufficient time for the air pressure in the compressor to build up significantly to 100 psi or more in both the secondary and primary systems.

(25) (B). The dual air brake is divided into primary and secondary systems.

The dual air brake system is divided into primary and secondary systems. The primary system controls the brakes at the rear axles while the secondary system takes care of the front axle, and sometimes, a rear axle. Both of these systems serve as air suppliers to the trailer.

35 Answers for Doubles and Triples Test

(1) (B). By using the parking brakes.

The parking brakes are designed to stop a vehicle from moving. When working on a double or triple trailer, apply the hand brake to keep the vehicle stationary until you are ready to move it. Alternatively, if the brake is faulty, chock the wheels to ensure the vehicle doesn't move.

(2) (A). Lowering the landing gear.

When uncoupling the converter dolly, the first step is to lower the landing gear. Then, disconnect the safety pins. This will give you access to other parts of the converter dolly you need to inspect.

(3) (D). When the dolly is under the rear trailer.

A good safety precaution when handling a double or triple trailer is to never detach or unlock the pintle hook when the dolly is under the rear trailer. If you unlock it in that condition, the trailer's weight makes the right hitch fly up and this can cause damage.

(4) (A). The first and second trailers.

The first and second trailers must be connected first when coupling a triple trailer.

(5) (C). Pull the dolly out and unhitch it.

When you are uncoupling your triple trailer, pull the dolly out first and unhitch it. After unhitching the dolly, you gain unrestricted access to the other parts of the vehicle on your checklist.

(6) (C). The lower section of the wheel.

The lower section of the fifth wheel should be checked first. The locking jaws and release arm are some of the areas you should inspect for possible damage or missing parts.

(7) (C). It is not properly lubricated.

The lower and upper sections of the fifth wheel are supposed to be close. If there are spaces between them, that's an indicator that the fifth wheel is not properly lubricated.

(8) (B). Check the manufacturer's handbook.

Sometimes, you may be stuck while coupling or uncoupling your trailer. When that happens, the first step is to check the manufacturer's handbook for advice.

(9) (D). None of the above.

When checking the lower part of the fifth wheel, examine the locking jaws, the release arm and the lubrication condition. Ensure there are no spaces between the lower and upper parts.

(10) (B). Trailer frame.

The glide plate should be mounted on the trailer frame. If there is an issue with the mounting, such as damaged or missing parts, this is a serious problem.

(11) (A). To avoid hitting each other during turns.

Always check the distance between the tractor frame and the fifth wheel to ensure that these two parts are not hitting each other during turns. When they are in constant contact, they may cause severe damage that can impact how the vehicle handles.

(12) (C). Fully raised.

The landing gear must be fully raised to enable it to function at maximum capacity. If it is not raised or is only partially raised, something has gone amiss and the landing gear may malfunction.

(13) (D). Both A and B.

Another name for the tractor air supply control is trailer emergency valve. It is also known as a tractor protection valve control. When pushed in, the vehicle supplies the tractor with the air pressure and the valve cuts the air off when pulled out.

(14) (D). None of the above.

A comprehensive inspection of the trailer emergency brakes should include its parts, air pressure and its emergency position. Each plays a different role in the overall performance of the brake system.

(15) (A). Chock the wheels or apply the brakes.

Trailers need a lot of air for their brake system. Thus, it is imperative that you check your vehicle's air pressure regularly. Before you start, apply the brakes. If that doesn't work, chock the wheels. Always ensure the vehicle is completely stationary during the test.

(16) (A). Supplying air to the service line.

If you can supply the air successfully, it means the airflow mechanism is in good condition.

(17) (B). Trailer service brake.

A trailer service brake is needed for intermittent testing. You can use it to slow down or stop the vehicle.

(18) (D). Both A and C.

The trailer service brake needs the air pressure to function and its reliability is a guarantee against brake failure.

(19) (C). A tractor protection valve.

The air brake system must be fully protected from damage. Ensure the tractor protection valve gets the amount of air pressure it needs. Otherwise, the brake system may lose air pressure.

(20) (C). 20 to 45 psi.

Although the trailer needs a regular air supply, airflow to the trailer should be stopped when the air pressure drops to between 20 and 40 psi. Do this before you disconnect the line.

(21) (C). Air leaks.

Some landing gear is power-operated. Hence, it must be thoroughly checked for possible issues. Air leaks will deprive the landing gear of what it needs to operate correctly.

(22) (C). By lowering the second trailer's landing gear.

The landing gear is a retractable support that comes in handy to support and stabilize a trailer when its tractor has been removed. Thus, it supports the second trailer and keeps it off the dolly, thereby reducing its load considerably.

(23) (A). Through the emergency line shutoff.

If you want to check the second trailer for air, you can do that via the emergency line shutoff. Open it to check the trailer's air supply. If there is no air in the trailer, its brake will malfunction.

(24) (C). The trailer.

The crack-the-whip effect is most pronounced in the trailer part of a combination vehicle. This usually occurs when a trailer changes lanes quickly. This can result in a turnover.

(25) (D). No.

Some states do not permit triple trailers. Hence, you should be fully aware of your state's stance on triple trailers before you apply for a CDL that will permit you to operate such vehicles. Be aware that several states are relaxing their stance against triple trailers so the laws could eventually change.

36 Answers for Hazmat Test

(1) (D). Proof of residency.

To transport hazardous materials, you need a valid CDL permit, a valid certificate from a medical examiner and proof of identity.

(2) (D). All of the above.

To prove your identity when applying for the endorsement, the testing center will accept your US passport, certificate of birth issued by a government-approved body, certificate of naturalization or certificate of US Citizenship. If you can't provide any of these, your identity can't be confirmed and your endorsement application will be rejected.

(3) (D). Three classes.

There are three classes of hazmat endorsement. These are the H Endorsement, X Endorsement and N Endorsement. Each endorsement allows you to drive under certain conditions.

(4) (C). N Endorsement.

If you wish to transport hazardous materials in a tanker, you need the N Endorsement on your CDL. The tank must be either temporarily or permanently attached to the chassis or the vehicle. Passing the Tank Knowledge Test is a requirement for getting this endorsement.

(5) (A). X Endorsement.

The X Endorsement is considered the ultimate endorsement class because with it, you can handle tanks and other hazardous materials. However, before you can obtain the X endorsement on your CDL, you must pass a test for hazardous materials as well as a skills and knowledge test for tankers.

(6) (A). $100.

The total cost of the hazmat endorsement on a driver's license is $100. The TSA charges $86.50 for screening and some others fees charged by the state bring the grand total to $100. None of these fees can be waived.

(7) (B). A new applicant with a valid TWIC card.

If you are a new applicant with a valid Transportation Worker Identification Credential card, you can enjoy the new reduced rate for such applicants: $67.

(8) (C). Transportation Worker Identification Credential.

A Transportation Worker Identification Credential (TWIC) gives the holder access to outer continental shelf facilities and port facilities that are regulated by the Maritime Transportation Security Act of 2002 with the primary objective of addressing port security across the country.

(9) (C). Five years.

If you are granted a hazmat endorsement after meeting all the requirements, the endorsement has a validity period of five years.

(10) (D). None of the above.

The TSA application requires your middle name, first name, date of birth, phone number and email address.

(11) (A). Phone number and email address.

Although you are required to provide a long list of useful pieces of information, phone number and email address are most important so that you can be contacted as needed.

(12) (D). Special characters.

The name field allows for letters, apostrophe, hyphens and spaces. However, special characters are not allowed. Failure to abide by this rule may invalidate your application.

(13) (C). No.

Applicants are not restricted to a specific location to take the test so long as it is within an applicant's state of residence. Visit the TSA's website and use the Enrollment Center Locator to choose where you take the assessment.

(14) (D). Your credit history.

To choose a convenient location, you must provide the following information: your zip code, city, the test you wish to enroll for and other pieces of relevant information. Your credit history will not be required.

(15) (D). 80%.

80% is the passing grade for the test. The section contains 30 questions and you are required to score a minimum of 24/30 to meet that percentage.

(16) (A). As many times as possible.

If you fail the test, you can take it as many times as necessary to pass it.

(17) (B). No.

You must pay all the necessary fees before being allowed to retake the test. However, you must wait for at least three days after the first test before submitting an application for a retake.

(18) (D). None of the above.

While taking the assessment, your knowledge in several areas will be tested. This includes your ability to identify hazardous materials, load shipments safely and placard your vehicle according to established rules.

(19) (D). Maritime Transportation Security Act of 2002.

A TWIC card grants its holders access to some vessels. These vessels are regulated by the Maritime Transportation Security Act of 2002. Among other things, the act was enacted by the US Congress to ensure safer roads for all users.

(20) (A). TSA charges plus state charges.

When applying for a hazmat endorsement, you are required to pay fees including TSA charges and state charges, totaling $100.

(21) (D). H Endorsement.

Drivers who transport placarded hazardous materials need an H Endorsement. It is imperative that you are familiar with both state and federal laws that guide the transportation of placarded hazardous materials and abide by them.

(22) (D). Medical Examiner's Certificate.

The DOT Card is otherwise known as the Medical Examiner's Certificate. This card is issued to drivers who pass the physical exam showing that they are physically fit to handle commercial vehicles for transporting hazardous materials. The medical certification process is overseen by the Federal Motor Carrier Safety Administration.

(23) (C). H Endorsement, X Endorsement and N Endorsement.

Hazardous endorsements are classified into the H Endorsement, X Endorsement and N Endorsement. An H Endorsement permits you to transport placarded hazardous materials while N and X Endorsements provide different permissions.

(24) (A). State government.

If the TSA grants your appeal or waiver, it will inform both you and the state government. This is to enable the state to be fully aware of your current status as well as the cancellation of your disqualification. Thus, you can then get the right endorsement, depending on your needs.

(25) (D). None of the above.

If a hazardous material contains an entry with an RQ or X in the appropriate column, you are handling a harmful material. A highlighted entry also indicates dangerous chemicals or materials. If you see any of these, you are transporting hazardous materials.

37 Answers for Tanker Test

(1) (A). Outage.

A space left on the top of a tank during loading to accommodate expansion is known as an outage. The allowance helps to prevent dangerous spillage due to the expansion of a tanker's contents.

(2) (C). Its expansion rate.

A very important attribute of a liquid each driver must know and take into consideration is its expansion rate. This enables the driver to know the appropriate space to leave for the liquid to expand without constituting a nuisance during transportation.

(3) (B). They are internal structures in a tank with some holes.

Although baffles are somewhat similar to bulkheads, they are nothing but internal structures and don't actually serve as dividers for tanks as bulkheads do.

(4) (B). Surge prevention.

Surge arises as a result of the impact of the liquid on the tank during transportation. The holes in baffles make them ideal for preventing surge that may occur when transporting liquids.

(5) (A). Baffles have internal partitions and smooth bore tanks don't.

While baffles have internal partitions that make them great for surge prevention, smooth bores don't. Thus, they serve different purposes from each other.

(6) (A). Smooth bore tanks.

The smooth bore tanks' design makes them ideal for transporting foods and food products. Thus, they are used for hauling fruit juice, milk, drinking water, wine, etc. Baffles can't serve the same purpose due to their different design.

(7) (D). Both A and C.

Driving on wet roads can be challenging. The absence of friction between the roads and tires makes the vehicle more prone to skidding, rollover and other accidents. Thus, when driving on wet roads, it is advisable that you increase your stopping distance as well as reduce your speed.

(8) (C). Remove your sunglasses.

When entering a tunnel, remove any sunglasses to enhance your vision in the dark. You should also be prepared for a gush of air that will meet you at the exit.

(9) (A). No.

When approaching a tunnel, you will likely see posted speed limits. These are only intended for smaller cars. Safe driving is your priority, not a specific speed limit. Make sure you drive slowly in a tunnel, regardless of whether the speed limit is higher.

(10) (C). The tanker driver.

If you accidentally spill your tanker's contents on the road, you are personally and legally responsible for cleaning up the mess.

(11) (A). Rollover.

Rollover occurs when the vehicle can't keep up with the steering speed. An effective preventive measure is to steer the vehicle gently to make it easier for the vehicle to respond accordingly.

(12) (A). Skidding.

Skidding occurs when a tire loses traction on the road. This can cause terrible accidents.

(13) (C). Release the brake immediately.

You may be inclined to keep the pressure on the brakes with the hope of stopping the vehicle during a skid. However, that is counterproductive. A more efficient solution is to release the brakes immediately. Then, try to control the vehicle as much as you can.

(14) (D). All of the above.

A leak in your vehicle can make driving difficult, if not dangerous. Thus, when inspecting your vehicle, check for leaks everywhere. Check under the vehicle for signs of leaking fluid. Check the tires as well. Don't leave anything unchecked.

(15) (A). Gaskets.

Ports and covers must always have gaskets. Otherwise, severe damage can occur to a vehicle. Ensure your inspection includes checking to make sure these gaskets are in place and undamaged.

(16) (D). All of the above.

A tanker driver who takes safety seriously should have some special purpose equipment that can guarantee his/her safety and that of the vehicle. These pieces of equipment include grounding or bonding cables, built-in fire extinguishers and vapor recovery kits, among others. They will come in handy during emergencies.

(17) (B). Bonding or grounding cables.

Electrical accessories are best connected to the vehicle's chassis with bonding or grounding cables. This prevents these accessories from getting dangerously near the engine and/or making unnecessary contacts that can trigger a serious reaction, especially when live cables come in contact with iron or anything else.

(18) (D). None of the above.

Apart from checking under the vehicle for leaks, you should check for other damaged or missing parts as well. This implies that your check should include the cut-off valves, hoses, vents, etc.

(19) (A). Between $10 and $50.

The Tanker Endorsement Knowledge Test is a special endorsement test for potential tanker drivers. The test will cost you between $10 and $50, depending on your state of residence.

(20) (C). Steady.

When you are applying the brakes, keep the pressure steady. This enables you to control the vehicle with ease and avoid abrupt braking that can cause a wide range of problems such as skidding and rollover.

(21) (c). Accelerate slightly.

It is advisable that you take some extra precautions when navigating bends. This is to ensure that you maintain the right speed for safety's sake.

(22) (D). Checking the mirror for incoming vehicles.

Skidding can be caused in tankers by an array of factors. The most prominent of these are overaccelerating, quick steering and excessive braking.

(23) (D). Manufacturer's manual.

Although Google and YouTube may offer some valuable tips, the most reliable source of information pertaining to your vehicle's maintenance is the manufacturer's manual.

(24) (B). To reduce pollution.

Vapor recovery kits are used for recovering gasoline vapors and preventing them from escaping into the atmosphere. The goal is to reduce both pollution and potentially explosive fumes. As a safety precaution, it is advisable that you have this kit in your vehicle to handle emergencies.

(25) (B). Personal Protective Equipment.

Personal Protective Equipment such as eye protection, gloves and hard hats can lessen your exposure to deadly hazards. Always include these pieces of equipment in your tool box because you don't know when you'll need them.

38 Answers for Passenger Transport Test

(1) (D). All of the above.

While checking under your vehicle, you should pay attention to several parts that include its frame, the exhaust system and the driveshaft. These parts play different roles in the vehicle and when any of them is faulty, it will have a huge impact on the vehicle's performance.

(2) (A). 5 miles/hour.

5 miles per hour is the most appropriate speed for testing the efficiency of the service brake. At that speed, it is easier to identify potential problems in the brake system as well as to gain insight into the overall efficiency of the system.

(3) (C). It depends on where a sticker is located.

A small sticker in the corner of the windshield where it doesn't obstruct the driver's view is harmless. The same can't be said of a large sticker that covers a significant portion of the driver's side, making it difficult for the driver to see clearly.

(4) (B). They are used to replace blown fuses or breakers.

Spare fuses or breakers are important vehicle accessories. They can blow out suddenly and that may pose a problem to the vehicle's electrical system.

(5) (B). Mounted and fully charged.

A fire extinguisher is of no use if it is not fully charged. Thus, you must ensure that your fire extinguisher is always fully charged and properly mounted where you can gain easy access to it during emergencies.

(6) (C). Red and reflective.

The red triangles are important safety equipment all drivers must have in their vehicles. They are used to warn drivers of a crash or accident ahead. For better visibility, especially at night, they must be red and reflective.

(7) (A). Clear and encourage free traffic flow.

The aisle is designed for free traffic flow, making movement easy for the passengers. Thus, it is imperative that aisles remain clear of obstructions.

(8) (C). Blind spots with the potential for triggering crashes.

No-Zones are blind spots along your driving route that may cause crashes or accidents if not properly identified and avoided. Running into a blind spot may cause a crash or accident that may damage the vehicle or vehicles, cause serious injuries, or loss of life.

(9) (C). Work areas and No-Zone areas.

Accidents can occur in work areas and No-Zone areas due to careless driving or excessive speeding. An effective preventive measure is to drive slowly while approaching such areas as well as when driving through them.

(10) (C). During the day.

While many drivers may be of the opinion that crashes are more frequent at night due to poor visibility, reports show that the reverse is the case because the frequency of road crashes during the day is higher.

(11) (C). Aggressive driving.

Some two-thirds of traffic casualties across the United States are caused by aggressive driving. This highlights the importance of driving defensively rather than aggressively.

(12) (A). Keeping a safe distance between vehicles and maintaining a safe speed.

Two of the major causes of road crashes and accidents are excessive speeding and driving too close to the vehicle in front. Many people drive too fast and leave little space to escape from danger.

(13) (C). Fastening your seat belt while driving.

Fastening your seat belt is arguably one of the most important safety precautions you can take while driving. The seat belt will hold you back during a crash, reducing the possibility of hitting the windshield with your head or hitting your chest on the steering wheel.

(14) (D). Small-arms ammunition.

While passenger buses are not designed for transporting ammunition, there are some exceptions. You can legally transport small arms ammunition that is clearly labeled ORM-D.

(15) (B). Obstruction of the doorway and damage to the baggage.

Leaving baggage in the doorway creates an obstruction to traffic. During emergencies or a stampede, the baggage may be destroyed completely or damaged beyond repair. Either of these situations can be avoided by keeping the doorway clear.

(16) (D). None of the above.

While passenger vehicles are allowed to carry some ammunition, drugs and emergency hospital supplies, under no condition are they permitted to carry division 2.3 poison gas, Class 6 poison or radioactive materials in spaces meant for human passengers. Failure to abide by these rules may result in severe penalties.

(17) (D). Liquid Class 6 poison.

You can transport ORM-D, emergency hospital supplies and small arms ammunition with your passenger vehicle. However, you are not licensed to transport Liquid Class 6 poisons such as phenol, clinical waste and some pesticides.

(18) (C). If they are for certified service or serve as guides.
Dogs and other animals are not allowed on a passenger vehicle. However, dogs can be allowed if they serve as guides for a physically challenged passenger or if they are certified service animals.

(19) (A). Embarking on a trip.
If you drive a charter vehicle, don't allow passengers to board your vehicle except when you are beginning a trip. This is to prevent vandalism to the vehicle and theft of passengers' baggage.

(20) (B). A set of triangles you can put beside the road as a warning signal to other drivers of a crash ahead.
A safety precaution for all drivers is to notify other road users of potential dangers. One way of doing this is to put three triangles beside the road to warn drivers of a crash or other danger ahead so that they can take necessary precautions.

(21) (B). A device used for connecting components of a drive train in a large vehicle.
A driveshaft, otherwise known as a tailshaft or propeller shaft, is a device used for connecting the components of a drive train in a large vehicle such as a passenger bus. When the shaft is faulty or damaged, it loses its connecting power and that may be disastrous.

(22) (D). All of the above.
It is very easy to detect a faulty driveshaft. The vehicle will shudder while accelerating, vibrate intensely in the undercarriage and/or be challenging to turn.

(23) (A). Cracks and loose bolts.
Loose bolts and cracks can cause a vehicle's frame to lose shape, raising the potential for an accident.

(24) (B). Failed or worn-out supporting bushings or bearings.
When the supportive bearing or bushing for the driveshaft U-joints or the driveshaft itself is failing, the result is usually clunking, rattling, squeaking or scraping.

(25) (D). None of the above.
Some common problems you may detect in your driveshaft include turning problems, shuddering during acceleration and excessive noise.

39 Answers for School Bus Test

(1) (A). Recognizing the problem.

When you are faced with an emergency situation, recognizing the problem is the first step towards finding a solution. When you identify the source of the emergency, you are better equipped to fix it.

(2) (A). Are they at risk of exposure to greater danger if they remain in the school bus?

When contemplating whether to evacuate the students during an emergency or not, one of the most important factors you should consider is whether students are safer in or out of the school bus.

(3) (D). All of the above.

While you may be considering whether evacuation is necessary or if there are better alternatives, some situations demand immediate evacuation. This includes if the bus changes positions or hazardous materials spill on or around it. Evacuation is also necessary if there is a threat of fire or accident.

(4) (B). Responsible and older students.

When planning ahead for emergencies that may demand student evacuation, you need the assistance of responsible and older students. You can delegate some responsibilities such as leading the students to a safe place while you handle other important tasks to make the evacuation faster.

(5) (A). Explaining the evacuation procedures to the students.

Prior to the evacuation, it is imperative that you explain the evacuation process to the students. This will help ease their fears.

(6) (C). Lead them upwind of the bus.

When students are at risk of exposure to spilled hazardous materials, leading them upwind of the bus is one way to help keep them safe.

(7) (B). Lead the students away from the tracks.

When you are faced with an emergency situation very close to railroad tracks, you must lead the students away from the tracks. Move them in the direction of oncoming trains where you have better chances of getting the needed assistance as soon as possible.

(8) (A). Shut the engine off and place the vehicle in neutral.

You shouldn't overlook the importance of securing your vehicle in an emergency. Secure your vehicle by placing the vehicle in neutral. Then set the parking brakes and shut the engine off. Remove the ignition key and activate the hazard lights.

(9) (B). Students with head or neck injuries.

Don't evacuate students with head or neck injuries until professional medical personnel are on the ground to assist.

(10) (D). None of the above.

When in danger or during emergencies, don't hesitate to ask passing motorists, pedestrians and responsible students for assistance. With their assistance, you can easily manage the crisis until a professional rescue team or other professionals are on the ground to offer professional assistance.

(11) (B). Accident management.

As a school bus driver, you should have a wide range of knowledge including accident management skills. Such skills will come in handy during emergencies such as crashes. This enables you to take care of yourself and your students before help arrives.

(12) (D). None of the above.

You shouldn't release the students to the neighbors, parents or other family members without being documented and attended to by medical professionals. This is to ensure that the students receive the best medical attention before they are handed over to their respective families. The school authorities have a say too.

(13) (D). A call to your supervisor or the school authorities.

If it is mandatory that you evacuate the students, call the school authorities and your supervisors too. They should be aware of each step you take. The school authorities should be able to account for their students as well. They may also send help your way.

(14) (D). All of the above.

Side doors, the roof hatch, rear door and front door are some exits you can use for the evacuation. Choose the safest and easiest of the available options.

(15) (B). Bus dispatch and the relevant authorities.

If you are ever in danger, steer clear of the bus. Wait for the relevant authorities or bus dispatch to declare it fit for use. In the meantime, do everything within your power to ensure that your passengers are kept safe outside the vehicle.

(16) (C). Evacuate the students to a safe assembly area.

Sometimes, the emergency situation may be caused by inclement weather that may affect visibility and make driving too risky. In that event, evacuate the students to a safe assembly area until the weather improves and driving is safer. Keeping the students in the vehicle may put your life and theirs at risk.

(17) (C). It refers to the act of barring people from exiting or entering a school until further notice in response to a potential danger within the school environment.

A school lockdown refers to the act of barring people from exiting or entering a school until further notice. A school may decide on a lockdown in the face of danger that may include things such as riots, shootings, internal crises, etc.

(18) (D). All of the above.

Apart from the potential causes of lockdown mentioned above, other common factors that can trigger a lockdown include police activity in the neighborhood, medical emergencies and natural disasters. The lockdown is a preventive measure against the potential threat and ensures the safety and security of the school, staff and students.

(19) (B). There are three types of lockdown.

There are three types of lockdown: Code Yellow, Code Green and Code Red. Each lockdown type is unique and is triggered by different forms of threats.

(20) (B). Code Yellow.

Medical emergencies, weather-related emergencies and traffic jams are some typical examples of situations that may trigger a Code Yellow. Although it signifies that danger isn't imminent within the school premises, all staff and students are still required to stay in the building.

(21) (C). Code Red.

Gunshots, a hostage situation and the presence of a dangerous person in or around the school are some situations that can trigger a Code Red. This situation may endanger everyone on the school premises. Thus, the students should be kept out of sight and this may require taking them to a safe place.

(22) (A). This occurs when the school bus is contaminated with hazardous materials.

A shelter-in-place situation may arise when the environment where the school bus is at a point in time is contaminated with hazardous materials that may endanger the lives of the passengers on the bus. You may either need to drive the vehicle away from the contaminated environment or seal off the bus completely.

(23) (C). All of the above.

Your vehicle may develop mechanical problems when least expected. To ensure the vehicle's safety as well as that of your passengers, it is advisable to drive to a safe place before stopping the vehicle. Then, secure the vehicle and activate the four-way flashers.

(24) (A). Electronic controller.

The electronic controller in the Antilock Brake System is designed to minimize skidding in school buses while helping the driver to maintain steering control even in extremely difficult braking conditions. It uses the wheel speed sensors to analyze the speed in all the vehicle's wheels and compare them.

(25) (A). Three times per permit.

You can retake the test within a couple of days or at your convenience. If you fail the first retake, you can only take the test again after waiting for at least 90 days to improve your knowledge and skills. Each permit allows you to retake the test three times.

Access the Bonus Audiobook and Flash Cards!

Scan the QR code below with your phone and you will be given a link where you can download the audiobook and flash cards:

If the QR code does not work for you, please contact us at info@newstonetestprep.com for the download link.

Newstone Test Prep